Masao Yoshida, Site Superintendent of the Fukushima Daiichi Nuclear Power Station, responding to reporters' questions with a grave expression on his face.

Units 1–4 at the Fukushima Daiichi Nuclear Power Station - TEPCO photo, March 16, 2011.

The Emergency Response Center at Fukushima Daiichi Nuclear Power Station three weeks after the accident. Site Superintendent Yoshida, center right and dressed in black, heads the response.

A long-boom pump truck called a giraffe pumping water into the nuclear reactors - TEPCO photo, March 23, 2011.

Fire engines amongst the scattered debris trying their best to pump water in. On the left can be seen a vehicle dispatched by the Self Defense Forces - TEPCO photo, March 16, 2011.

On site at the Fukushima Daiichi Nuclear Power Station damaged by the Great Eastern Japan Earthquake tsunami - TEPCO photo, March 11, 2011.

Checking onsite safety in protection suits some four years after the accident. Holding tanks for contaminated water can be seen in the background
- TEPCO photo, January 21, 2015.

The spent fuel pool at Unit 4
- March 12, 2014.

The Anti-Seismic Building that acted as the accident response headquarters - March 12, 2014

A thousand paper cranes hung in the stairwell of the Anti-Seismic Building to bolster workers' spirits.

The entrance to the Anti-Seismic Building. The words, "We're counting on you, Fukushima Daiichi!" catch the eye.

An aerial shot prior to the accident showing the six Units at the Fukushima Daiichi Nuclear Power Station. The huge tsunami attacked from the foreground, extensively damaging the major facilities by either flooding or destroying them - TEPCO photo

Anatomy of the Yoshida Testimony

―――The Fukushima Nuclear Crisis
as seen through the Yoshida Hearings

Rebuild Japan Initiative Foundation

Anatomy of the Yoshida Testimony
—— The Fukushima Nuclear Crisis as seen through the Yoshida Hearings

First edition, August 28, 2015.

Author	Rebuild Japan Initiative Foundation. 11F Ark Hills Front Tower RoP 2-23-1 Akasaka, Minato-ku, Tokyo 107-0052, Japan Phone +81-3-5545-6733 Fax +81-3-5545-6744 URL http://rebuildjpn.org/en/
Publisher	Shuzo Tanabe
Publishing	Toyo Publishing Co., Ltd. 1-23-6 Sekiguchi, Bunkyo-ku, Tokyo 112-0014, Japan Phone +81-3-5261-1004 URL http://www.toyo-shuppan.com/
Art Direction	Masato Kumazawa
Design/Layout	Akiyoshi Murana (Power House Co., Ltd.)
Illustration	Sachiko Maruyama
Proofreader	Kaori Maruyama (Ensuisha Co., Ltd.)
Printing	Nihon Hicom Co., Ltd.

© 2015 Rebuild Japan Initiative Foundation.
This publication is in copyright. Subject to statutory exception
and to the provisions of relevant collective licensing agreements,
no reproduction of any part may take place without
the written prermission of Rebuild Japan Initiative Foundation.

All the information available at the URLs below is
in Japanese and was valid as of August 18th, 2015

ISBN 978-4-8096-7801-1

TABLE OF CONTENTS

Introduction (Glossary) 1 4
Introduction (Glossary) 2 5
Introduction (Glossary) 3 7

Foreword: **From Investigating the Accident to Investigating the Crisis** 9

[PART 1] **A Reading of the Yoshida Hearings** 15
Yoichi FUNABASHI

[PART 2] **Three Years on from Publishing The Independent Investigation Report** 87
(An analysis and investigation three years after the event by four members of the then Working Committee)

Kenta HORIO 88
Akihisa SHIOZAKI 97
Kazuto SUZUKI 109
Shinetsu SUGAWARA 119

Afterword 127
Project Members 132
Chronology of the Fukushima Nuclear Accident 134

The Rebuild Japan Initiative Foundation is an independent think tank established in September 2011, investigating the cause of the Fukushima nuclear accident being its first subject.

Under the motto of "Truth, Independence, Humanity", it researches and provides policy proposals on national issues facing Japan. It hopes to engage in issues with citizens all over the world by publishing its reports in English. Its English publications to date include The Fukushima Daiichi Nuclear Power Station Disaster, and Quiet Deterrence.

GLOSSARY 1

The Fukushima Nuclear Accident

The nuclear accident at TEPCO's Fukushima Daiichi Nuclear Power Station, which was triggered by the earthquake and tsunami caused by the Great East Japan Earthquake that took place at 14:46, March 11, 2011. At the time of the earthquake, Units 1-3 were in operation; Unit 4 was undergoing inspection; and Units 5 and 6 were undergoing a regular inspection. Units 1-3 stopped automatically in an emergency shutdown, but the entire external power supply was cut off due to extensive damage to electrical equipment caused by the strong tremors, including the collapse of power plant towers. The emergency diesel generator in place in case of a loss of external power supply started automatically, but some 50 minutes after the earthquake, the plant was struck by a huge 15-meter high tsunami, and apart from the diesel generator installed in a building basement shutting down, electrical equipment including the power panel were inundated and could no longer be used, the power plant falling into an emergency state of total power loss.

As a result, the enormous heat generated in the reactors and the spent fuel pools could no longer be cooled in Units 1-3, and with each passing hour, nuclear fuel damaged the reactor core, molten nuclear fuel dropped to the bottom of the pressure vessel, and the molten fuel leaked out of the containment vessel in a melt-through. In addition, explosions caused by the mass generation of hydrogen inside the containment vessel took place in Unit 1 sometime after 15:00 on March 12 and in Unit 3 around 11:00 on March 14.

In order to prevent pressure in the reactor containment vessel from climbing too high, emergency venting was carried out to reduce the pressure by discharging gas containing radioactive material into the atmosphere, and this combined with the damage to equipment caused by the explosions led to a large amount of radioactive material being released, creating a serious accident classified as a Level 7, the highest level on the International Atomic Energy Agency (IAEA) International Nuclear Event Scale

(INES), placing it on a par with the former Soviet Union Chernobyl nuclear power plant accident of 1986.

In order to avoid damage caused by radioactive material, the Government issued an evacuation order on the night of March 11 for all residents within a 3-kilometer radius of the Daiichi Nuclear Station, expanding this to 20 kilometers on March 12. Evacuees were estimated at around 17 million people, with some 100,000 inhabitants still in evacuation four years after the accident. Although there were no direct deaths linked to the nuclear accident uunlike Chernobyl, it was a disaster of unprecedented proportions with more than a thousand deaths among hospital patients and nursing home residents deemed to be related to the emergency evacuation, which took place in the midwinter cold.

GLOSSARY 2

The Yoshida Hearings

The Yoshida Hearings refers to the record of interviews by the Government Investigation Committee on the Accident at the Fukushima Nuclear Power Stations with TEPCO Fukushima Daiichi Nuclear Power Station Site Superintendent Masao Yoshida, who led the emergency response. They were conducted a total of thirteen times from July to November 2011 in a Q&A format running for some 29 hours. The hearings are an extensive account exceeding more than 300 pages, which the Cabinet Secretariat has published on the web.

A public prosecutor seconded to the Government Investigation Committee was mainly in charge of the hearings. Initially the investigation was not made public due to a request from Superintendent Yoshida to "not go public because of the risk of factual errors owing to fading and/or confusion of memory", the Asahi Shimbun being the first to cover it in May 2014. Since other major newspapers including the Sankei Shimbun and the Yomuri Shimbun followed suit, the Government went public with the report on September 11.

Yoshida was a nuclear engineer with a Master's Degree from the Tokyo Institute of Technology, and this was his fourth posting at Fukushima Daiichi Nuclear Power Station. Well familiar with the plant's facilities and equipment, he assumed command of the response in the immediate aftermath of the accident working night and day. In November, 8 months after the accident, he was hospitalized with esophageal cancer, and was in and out of hospital until his death in July 2013. He was 58 years old.

The Government Investigation Committee is the only investigation Yoshida provided a full hearing to, and although the National Diet Investigation Commission did interview him, it was unable to perform its own hearing, compiling its report on the basis of the Government hearing results. The Yoshida Hearings are a vivid personal account of the turmoil and confusion at the plant immediately following the earthquake, the miscommunication and lack of communication between Fukushima Daiichi NPS and TEPCO Head Office and the prime ministerial Kantei, and Yoshida's own resolve to die at one stage as the nuclear reactor situation deteriorated. As a candid record from the person in charge, it is a valuable source for evaluating and analyzing the accident.

The Government subsequently went public with the records of hearings with more than 200 people, including Kan Naoto, the Prime Minister, other politicians and those involved in the response at the power plants as well as subcontractors, but the hearing results of top executives at TEPCO Head Office at the time of the accident remain unpublished.

GLOSSARY 3

The Independent Investigation Commission

(The Independent Investigation Commission on the Fukushima Nuclear Accident)

The Rebuild Japan Initiative Foundation (Chairman: Funabashi Yoichi) was established six months after the Great East Japan Earthquake in September 2011 in order to investigate and analyze the cause of the Fukushima nuclear accident and the crisis response at the time. Chaired by Kitazawa Koichi, former Chairman of the Japan Science and Technology Agency, the Commission comprised six renowned experts in the fields of nuclear power, law, governance, international relations and so on. The actual investigation process was carried out under the guidance and supervision of the Commission by a working group of some 30 lawyers, journalists and researchers in nuclear power and international politics, with interviews from more than 300 people undergoing diverse analysis. A 400-page long report was published immediately prior to the first anniversary of the earthquake at the end of February, 2012.

The main organizations to have investigated the Fukushima nuclear accident are the Government Investigation Committee (chaired by Yotaro Hatamura), and the Diet Investigation Commission (chaired by Kurokawa Kiyoshi), both of which had a certain enforceability. A purely private organization such as the Independent Commission had to rely on the cooperation of stakeholders and was not able to secure any substantial cooperation from TEPCO, the party directly involved in the accident. On the other hand, it received a great deal of voluntary cooperation from parties who agreed with the Commission's esprit of political neutrality, its lack of organizational ties and its goal of an independent investigation. Politicians in charge of the response at the time including the then Prime Minister Kan, Chief Cabinet Secretary Yukio Edano, METI Minister Banri Kaieda, nuclear experts such as Haruki Madarame, Chairman of the Nuclear Safety Commission, and Shunsuke Kondo Chairman of the Japan Atomic Energy Commission, officers and officials directly and indirectly involved in the response both in the field and in government institutions also cooperated with the investigation, as did even some ex-TEPCO employees. The politicians' hearings have already been published on the web.

The Independent Investigation Commission compiled its report ahead of those of the Government and Diet Investigations, and was the first to point out the confusion triggered by politicians including excessive interference from the Kantei and the "nuclear safety myth" that "a severe nuclear accident would never happen in Japan" that acted as a backdrop to the accident. It also placed the nuclear accident in the structural and historical context of nuclear development in Japan, as well as discussed in the U.S.-Japan relationship over the nuclear accident, which had formerly not been given much attention, attracting a great response not only from the domestic media but foreign media as well.

The Japanese report was made available from Discover - Twenty One, Inc. in March 2012, with an English version based on the Japanese version also being published by Routledge in 2014.

Japanese Version

English Version

Independent Investigation Commission Website
http://rebuildjpn.org/project/fukushima/

Foreword:
From Accident Investigation to Crisis Investigation

The Government Investigation Committee hearings with Masao Yoshida, Site Superintendent of the TEPCO Fukushima Daiichi Nuclear Power Station when the Fukushima nuclear accident took place (Yoshida Hearings), provide a wealth of pointers on why a crisis came into being; why the crisis could not be prevented; what was the historical, structural background; and what lessons should be learnt.

All things being equal, the Yoshida Hearings may never have seen the light of day. In fact, Yoshida himself responded to the Government Investigation in the belief that its findings would not be made available to the general public.

Yoshida met with members of the Government Investigation in July, August, October, and November of 2011, as well as with the Diet Investigation in May 2012 when he was hospitalized. At the time, Yoshida agreed to the Diet Investigation side's request, signing a document to the effect that he "did not object to disclosure" of the Yoshida Hearings drawn up by the Government Investigation in the Diet Investigation report.

However, this disclosure referred merely to contributing to the internal investigations of the Diet Investigation, Yoshida having clearly petitioned the Government Investigation (on May 29 of that year) that he did not wish his statement to be made available to the general public. He also requested the Diet Investigation Commission to exercise strict control so that the material would not to be leaked to a third party.

Yoshida was concerned that the "sections where I frankly stated my evaluation of others, feelings, and impressions", "could lead to misunderstandings if not considered in the context or in the light of my psychological state and the atmosphere at the time of the hearing, as well as my relationship with the officer conducting the hearing," as well as the possibility that he may have "mistaken the facts" owing to "a fading memory associated with the passage of time, as well as confusion of memory due to the fact that I was forced to deal in quick succession with various events." Yoshida confessed to the concern that he was frightened that "Everything I said might be taken as a matter of fact and assume a life of its own."

That his statement has been made public in this way must, therefore, go against his will completely.

There has been an attempt in the form of Truth and Reconciliation Commissions (TRC) to examine as well as make public the causes and background to genocide and systemic ethnic and racial discrimination so that it may never be repeated. The approach adopted here is to have people speak the truth by undertaking not to disclose the results of hearings or to pursue the individual criminal responsibility of the parties involved in order to uncover the truth. There is also a growing trend to uncover the truth about transportation accidents by similarly not questioning criminal responsibility. Regarding the Fukushima nuclear accident, it would not be wrong to regard the Government Investigation as a so-called Truth and Lessons Commission (TLC) and to follow such an approach.

Yoshida was quite insistent that disclosure to the general public should not take place. From this perspective, the government would appear to have taken advantage of Yoshida's death and betrayed him. An unresolved issued especially is the fact that the government provided no reasons for its actions when it placed the Yoshida Hearings in the public domain.

There was a growing demand for the entire text of the Yoshida Hearings to be made public after it was partially reported on, which led to questions concerning the accuracy and fairness of the coverage itself. Although its release by the government can be seen as responding to this demand, the government's intent and rationale were ambiguous.

Investigating the cause of and response to the Fukushima nuclear accident is an increasingly important issue today. Uncovering the truth and learning the lessons it provides is essential for raising both the safety of nuclear power plants and improving a culture of safety. This being so, it should have been possible to reason that the full disclosure, albeit an exception to the rule, would contribute to the greater public good. And, whether or not it was in the public interest depends on whether or not we put it to that use. That is the awareness that has driven our attempt here.

This is how the Yoshida Hearings resulted in being the vivid and powerfully uncensored account it is. The voice in this record is the real voice of the now deceased Yoshida.

Sometimes he is intense, other times subdued; sometimes on the defensive, other times contrite. And in certain places, there are inconsistencies and confusion.

And more than anything else, it is the product of memory. Being a human memory, it may be selective and there may be mistakes. Nevertheless, or rather because of that, the hearings are a valuable document for understanding the truth of the Fukushima crisis. It is the timbre of a human voice that gives this framework of truth a three-dimensionality, or rather a physicality.

It is not possible to understand the essence of a human social crisis, especially something like a severe nuclear accident, which is also inevitably a national crisis, without taking into account the human factors that weave together the crisis management elements of risk, governance, leadership, and organizational culture. In short, it is not possible to capture its essence. The more severe the crisis, the more decisive these human factors become. In his statement, Yoshida confesses and exposes his entire being, providing an irreplaceable insight into these human factors.

This record is Masao Yoshida's testimony.

We should accept it as such, listen to his voice, grasp the truth lurking there, and draw the lessons it imbibes.

The Rebuild Japan Initiative Foundation established an independent investigation (The Independent Investigation Commission on the Fukushima Nuclear Accident) in the wake of the accident to investigate and analyze the causes and background under the motto of "Truth, Independence and Humanity", releasing its report in February, 2012. An English version was subsequently published by the British publisher Routledge in 2014 under the title, *The Fukushima Daiichi Nuclear Power Station Disaster Investigating the Myth and Reality.*

The Independent Investigation did not confine itself to investigating the accident alone, but was committed to highlighting the essential nature of the crisis behind the accident and the response by focusing on the risk, governance, leadership and organizational culture involved in that response. Its sights were set on not just the accident, but on investigating the crisis.

This current attempt to decipher the Yoshida Hearings is founded on the same intent. However, it is not part of the Independent Investigation Commission, which was disbanded after its final report was released. However, investigating the essence of the Fukushima nuclear crisis remains a major theme for the Rebuild Japan Initiative Foundation.

This publication is part of that process. Specifically, four members of the Independent Investigation Commission's Working Group (Kenta Horio, Kazuto Suzuki, Shinetsu Sugawara, Akihisa Shiozaki) played a central role in examining the Yoshida Hearings, and in the light of it, have evaluated facts uncovered by the Independent Investigation's report and its interpretation. Each has acted in a personal capacity rather than as a representative of their respective institutions.

February, 2015
Yoichi FUNABASHI
(Chairman, Rebuild Japan Initiative Foundation)

[PART 1]

A Reading of the Yoshida Hearings

Yoichi FUNABASHI

I   The "Unanticipated" is the Greatest Enemy of Crisis Management

> *Q: In the end, what sort of response did you think of taking when power needed to be restored to the instruments and you were unable to use the emergency diesel?*
>
> *A: I myself was basically in despair. We were entering a severe accident (...), but after they stopped, after the batteries stopped, we were saying we needed to consider how we were going to do the cooling, but I couldn't come up with anything.*
>
> *Q: There was no answer.*
>
> *A: There was no answer*
>
> The Yoshida Hearings, July 22, 2011

The crisis began at 15:37, March 11, 2011.

The shift supervisor for Unit 1 reported to the Roundtable in the Emergency Response Center (ERC) on the second floor of the Anti-Seismic Building that "All the emergency diesel generators in Unit 1 have stopped. We've lost all power!" These were the diesel-fired emergency generators (D/G).

The external power supply had already stopped.

The loss of all power, even the emergency supply, meant nothing less than the entire loss of power to cool the reactors.

Yoshida testified what he felt at the time in the following manner.

"I was devastated. (...) It was a real mess. There was a strong possibility of it becoming [a severe accident]. We have to prepare for that. Can't we get the D/G firing? (...) What are we going to do if we lose them? If we had an isolation condenser (IC) or an RCIC, we could keep cooling for several hours, but what are we going to do next? This was what kept going around and around in my mind (July 22).

The IC (emergency condenser) was an emergency reactor cooling facility located in Unit 1. The extremely hot steam generated in the reactor was cooled by running it through coiled pipes in the cooling tank of the IC on the fourth floor of the reactor building, then sending it back to the furnace as water. There was also a reactor core isolation cooling system (RCIC) where the reactor steam turned a turbine-driven pump that injected cooling water into the reactor in Unit 2.

In the case of a loss of both external power supplies and emergency power supplies, namely an SBO (station blackout), these cooling systems were to go into operation and cool the reactor core.

Since these cooling systems use DC power, the reactor core can be cooled without an AC power supply. However, the batteries in Units 1 and 2 were both flooded by the tsunami and the DC power supply was lost.

In its Safety Design Review Guidelines for Power Generation by Light-Water Nuclear Reactor Facilities (drawn up August 30, 1990), the Nuclear Safety Commission wrote "reactor facilities are designed to safely shut down the reactor core and maintain cooling after shutdown for short SBOs", adding its endorsement that "there is no need to consider prolonged SBOs because restoration of power transmission lines and/or emergency AC power supply equipment can be expected."

It deemed a "prolonged SBO" to be "unanticipated". And given that it was "unanticipated", no preparations were made. In fact, not to prepare, in short, to remain vulnerable in such a case was taken as a "proof of safety".

An SBO in a nuclear accident disaster was more or less expected – but only for a short period of time. Training had also been carried out on how to respond to such an event. However, a loss of the DC power supply did not even figure. The loss of the DC power supply was "unanticipated in the unanticipated". If it was lost, neither the IC nor the RCIC would function.

One of the engineers on duty at the Roundtable recalled to me, "Perhaps the AC power supply, but no-one imagined a scenario where the DC would be lost, not just Yoshida-san."

Yoshida's mission at the time was to cool five reactors (of the six reactors, Unit 4 was in the process of having its shroud, a cylindrical structure that housed the reactor pressure vessel, replaced). Where should he start?

Should he connect the batteries to the instruments for Unit 1 or Unit 2?
Should he start venting from Unit 1 or Unit 2?
Units 1 and Unit 2 were operated from the same Central Control Room.
The accident developed simultaneously and in parallel into a "parallel chain nuclear disaster". This made the response difficult.

Yoshida's second-in-command commented on this point to me as follows.

"We didn't know at the time if we were running a 100-meter sprint, a marathon, or an ultra marathon. The crises came in one by one, Units 1, 2, and 3. That helped us. Yoshida-san also said later that that was what saved us."

Meltdown occurred one by one in Units 1, 3, and 2. They did not take place on the same day at almost exactly the same time. It was this turn of events that was expressed as "what saved us".

In a crisis, the most difficult thing to realize is a sense of "sequence" and "order".
Conversely, to lose these is a crisis.

- What is happening now?
- What will happen if no action is taken?
- What new crisis may be triggered by this crisis?
- What needs to be done and by when to prevent the crisis from escalating, to overcome it?

This is the situational awareness required in crisis management. Positing the loss of DC power supply as "unanticipated" made any such situational awareness extremely difficult.

Yoshida expressed contrition in the hearings, saying, "Of late, every time I use the word 'unanticipated', I get beat up so it's difficult to use, but it's not an issue you can escape from by saying it's an operational thing. You enter a kind of mental block. (November 6)."

Other examples that led to severe nuclear accident include the 1979 Three Mile Island (TMI) nuclear accident in the United States and the 1986 Chernobyl nuclear accident in the former Soviet Union.

Every time these cases occur, understanding of and responses to severe accidents in the countries of the world that use nuclear power have moved forward, and measures have been put in place including accident management (AM). This is so in situations of extensive damage beyond the design basis events for reactor cores (severe accidents), and to prevent expansion or to mitigate the impact in the case of expansion.

Japan, however, did not make accident management a regulatory requirement, nor did it try to expand its scope. What a defenceless and irresponsible regime.
Yoshida learnt this on that day.

One of the TEPCO nuclear executives reading the Yoshida Hearings let the impression slip that "defeat was already in our sights when the battle began."

He recollected, "I have long entertained doubts as to why TEPCO was so useless with all the engineers it had. Reading the Hearings brought it home to me anew. Without preparation, you're helpless. There were multiple reactors simultaneously headed into crisis in parallel. It was impossible no matter what you did if you didn't have the training."

The nuclear defeat was preordained by creating the "unanticipated". This was because it gave birth to a "mental block".
The "unanticipated" is the greatest enemy of crisis management.

II Measurement is the Essence of Crisis Management

> *A: I had never used that kind of IC (...) and I myself had very little understanding of how to control the IC itself. I had been the repair manager for Units 1 and 2, so in the sense of repairs, operationally speaking, you open the IC so much, the steam comes out and is condensed then returned, so the water level fluctuates, but I have no idea. Sorry.*
>
> *Q: I mean when the AC power supply was lost, or even immediately after the loss, did you think that the IC was operating then?*
>
> *A: Yes*
>
> (August 8)
>
> *A: Thinking about it now, it was a mistake to believe to a certain extent the water level meter. I deeply regret having placed too much faith in [the water level shown by the water level meter].*
>
> (July 22)

Part 1 A Reading of the Yoshida Hearings

After the reactor shutdown, the IC started up automatically, but the inundation caused by the tsunami cut off all the power, both AC and DC. The lights that repeatedly switched on and off indicating by color whether the valves were open or shut to prevent an abrupt cooling of the reactor went out, meaning there was no way to tell if the IC was operating or not.

However, Yoshida responded in the hearings that he thought the IC was operating all the time. Since he thought that the water level in the reactor was being maintained, he said he was sure the IC was functioning.

Still, by the time the decision to vent was made at midnight on the 12th he stated he had his doubts as to whether the IC was functioning or not. In fact, the IC had already stopped by that time.

"I also thought something was funny (…) it's written at 21:51, and that was the dose. Why is the dose climbing so high (…) the IC is working, right? The water level is on the plus side, but the dose is up, that's strange (…) that's about when I started having doubts. So, although it says we were only checking the water level, I started thinking something strange was going on, that perhaps the IC had stopped, in other words that we had lost our cooling source. (July 22)"

What these statements attest to is that Yoshida (and not only him, but many other workers responding to the accident) did not know enough about the role and design of the IC. And with only an ambiguous understanding of whether it was working or not, they failed to confirm.

One of the officers at the Roundtable with Yoshida told me, "My gut reaction was that it was barely working. I didn't think it was working solidly. When we got the TAF figure of 200 millimetres, I felt it wasn't in good shape. After all, 200 millimetres is 20 centimetres."

He was speaking of the shock he remembered when the water

level in the pressure vessel was a mere 20 centimetres above the Top of Active Fuel (TAF). This was after 9p.m. on the 11th.

The IC was the so-called "lifeline" in the case of a total AC power outage. Nevertheless, at least reading the Hearings, you do not get the sense that that was how Yoshida perceived it.

This mistaken situational awareness not only derives from "being convinced" that the water level meter was probably functioning, but also a lack of smooth communication between the Central Control Room and the Emergency Response Center as well as between Yoshida and his team leaders around the Roundtable.

At the risk of repetition, Japanese nuclear safety regulations deemed a severe accident entailing a prolonged loss of all AC power as "unanticipated", and the loss of the DC power supply loss as "unanticipated in the unanticipated".

So, when such an accident occurred, the expected role and actions of the site superintendent were unclear. This is where the essence of the problem lay.

A former TEPCO nuclear engineer, Professor Omoto Akira of the Tokyo Institute of Technology, has pointed out the problems in the IC response as follows.

"There is the issue that the actual design [of the IC] was not well understood. They didn't know about the large volume of vapor released into the atmosphere when the IC was running. It's fairly common knowledge that the water level meter can no longer be relied on when the temperature rises in the containment vessel of a BWR [boiling water reactor like Fukushima Daichi], so it's quite surprising to think that no steps were taken to deal with this."

"Also, no objections have been raised regarding the misjudgement that 'the IC in Unit 1 is working, so that's OK'. It is clear from the Hearings that there was no group in place gathering information from each functional team, integrating and calmly analysing it, then advising the site superintendent."

(Omoto Akira, On reading the Yoshida Hearings, personal mail to Funabashi, November 28, 2014)

However, there may be another human psychological factor at work here wanting to reject adverse information. Yoshida states the following in the hearings.

"This is a point of personal regret, but although there was a conviction, no information from the power generation group leader (whether or not the IC was working) was received at the Roundtable. So, I don't know whether information was reaching the power generation group leader from the shift supervisor or not. (...) Since things weren't set up so that the shift supervisor could phone me, I should have realized we needed to check over and over if the IC was OK. (July 22)"

At the same time, Yoshida's regret that he "placed too much faith in the water level shown by the water level meter" also seems strange.

The fact that the water level meter and the water level is tricky in the case of a total power outage was one of the lessons learned from the Three Mile Island accident, and the movie based on it, The China Syndrome (1979, directed by James Bridges), also dramatized the lack of reliability of the water level as indicated by the water level meter.

While operators in the field from a very early stage were questioning, "you can't trust this water level", their awareness was not shared by the Roundtable. It seems unfathomable that Yoshida and other top managers had no reflexive reaction that the water level figures were chancy.

One of Yoshida's "lieutenants" confessed to me, "Being totally in the dark, the only raw data we were getting was the water level. In the absence of any other yardstick, the numbers seemed real."

Indeed, the numbers shown by water level meter were wrong, but that doesn't mean they were deliberately wrong. The numbers were strange because the water level meter had broken down. A TEPCO executive from the nuclear division had the following to say.

"I thought the instrument was correct. This time, not many of the instruments were deceptive. Even when the pressure fell in the containment vessel on the 14th, the instrumentation wasn't showing zero, but more that it had had it, in other words, it was showing that it wasn't working, not that the pressure was at zero. Read that way, the instruments were telling the truth."

Airplane pilots undergo recovery drills from quasi crash scenarios. At that time, the human senses are destroyed and they lose all sense of up and down, and so they are taught to believe only in the instruments and make the recovery timing their steps. That is how they are trained.

Yoshida's "deep regret" should be seen as not being able to maintain the instrumentation and scales during the crisis. This is because without instrumentation as a reference, it is not possible to grasp the essence or direction, the scale of the situation or how things are progressing.

In a crisis, the instrumentation must be protected at all costs because they form the basis of measurement. In a crisis, what can be measured can be managed.
Measurement is the essence of crisis management.

III The Strategic Issue of Logistics

A: As to whether the materials team people knew the specifications, they didn't. The materials team (...) don't know

> the fine technical specifications, and so the recovery team had to provide the specifications. For example, they had to specify what voltage batteries, something like this, or how many kW power vehicles, which was quite difficult.
>
> We had to send to the Onahama Coal Center, our base in Onahama (Iwaki City, Fukushima Prefecture) and bring things out from there, but because the dose was climbing, things couldn't be brought in. In short, because there was no means of transport, we were told to come and get it ourselves and in the end logistics were a problem.
>
> Our strong demand was that things that were just the right fit were brought to us.
>
> (July 22)

During the Fukushima nuclear accident, TEPCO was having trouble with this "logistics problem".

Yoshida repeatedly asked Head Office to "send us stacks" of "water, gasoline and diesel", but the response of Head Office was bureaucratic and makeshift, and they were unable to provide the pinpoint response of "just the right fit here", everything being half-hearted.

A messy state of affairs reigned with the gasoline sent at the request of Fukushima Daiichi turning into "a total fiasco with it being unloaded at Fukushima Daini because they want gasoline as well".

Moreover, no drivers were willing to go with the local radiation dose increasing. TEPCO had not set up in advance a transport system at the time of such a crisis.

People in the field responding to the crisis including Yoshida set up a command tower in the Anti-Seismic Building. They needed kerosene to keep the emergency power running there, but the kerosene tanks were damaged by the tsunami. If the kerosene ran out, "the entire chain of command would come to a halt." Head Office did not share this sense of crisis. Irritation with this was growing at the site.

It was obvious at a glance that there were no material preparations, be it power vehicles, batteries or gasoline. However, in the case of events and processes, this was difficult to understand.

What would happen if the accident's integrated chain of command (ICS) was "unprepared"? For example, whose responsibility was it to keep an eye on the water level of the pressure vessel?

Yoshida stated in the hearings, "Who was to check the water level, the fire brigade or the maintenance team? Or cool the water? Who was to go into the field and check? Everything was in confusion (November 6)."

Even during the brink strategy to open the SR valve (the main steam safety relief valve) in Unit 2 on the evening of the 14th, this lack of preparation led to confusion.

"Was it the operators who were to open the SR valve or the maintenance personnel? People from power generation asked why they were doing it and I was forced to say things like don't be stupid, just do it. After all, in such a state, I had to challenge them (November 6)."

What of the "preparations" for communication?
Despite a video link between Head Office and the site, communication did not go well.

"There were many queries. What's the situation now and so on. That wasn't support. They were just asking so they could report back. I lost my temper halfway through and remember telling them to pipe down or shut up more than once (July 29)."

These types of ICS "inquiries" that were not for "support" were observed not only between Head Office and the Roundtable, but also between the Roundtable and the Central Control Room.

"It was a mere 100 meters to the site, and they kept contacting us saying isn't there something that can be done (…) or can't you do more, but actually it wasn't possible to do anything where we were. This gap lasted the whole time (July 22)."

By way of supplementing Yoshida's statement, one of his "lieutenants" has stated that "there was a communication problem with Head Office, but communication inside the plant, I was keenly aware that that was the real core."

Concrete response to an accident is firstly the responsibility of the shift supervisor in the Central Control Room. To the shift supervisor, sending workers into the vicinity of the reactor was akin to sending in a "death squad" as the fear of aftershocks and tsunami rose as well as the risk of a higher dose, a growing danger of an explosion and the fear of exposure. Confirmation operations proceeded slowly.

Yoshida, although disappointed there was no SOS from the Central Control Room, regretted even more that he didn't offer them help.

Communication becomes difficult because normal means of communication often cannot be used in a crisis. In such a case, it is no longer possible to share information, evaluations and judgment calls during the response decision-making.

It is not only information that needs to be shared.
Psychological support in the form of whether there is a shared understanding of the situation the other party is in is another key element of sharing. Whether at Head Office or in the Emergency Response Center, assistance needs to take into consideration this kind of psychological support.

Yoshida has stated, "When it comes to specific operational work, in a sense, the operator is the professional, so things are left up to the power generation group leader and the operators. The detailed operational procedures, that is (July 22)."

The Yoshida who emerges as site superintendent from the hearing records is certainly not a hands-on craftsman hammering each nail or planing each board, but more of a foreman sent to a building site by a construction company. Rather, it is true that a site superintendent intent on micro-management can create problems. However, regarding "lifelines" such as the IC or emergency responses such as venting, the site superintendent has to know the subtleties involved. This is true for normal operations, but even more so in an emergency. In "unanticipated" circumstances he has to deal with the situation by using his imagination for when operational procedures are useless whether the issue be "small and detailed" or a "substantial" one. Things cannot just be "left up" to those in the field. He must be thoroughly cooperative with the site.

Yoshida felt isolated due to the lack of support from Head Office. However, the shift supervisors must have also been feeling isolated with the lack of support from the Emergency Response Center. A TEPCO engineer recalls that the shift supervisor at Unit 3 must have acted under just such circumstances when he shut down the HPCI (high pressure water injection system).

"The shift supervisor was doing his best in isolation at the time with no consultation. If I had been asked at the time, Î would have immediately advised against doing it. It's absolutely imperative not to stop the HPCI. Keep it running even if it bursts. That's what I would have advised. Instead, it was like he shut it down somehow, yes, I see."

One of the nuclear power executives at TEPCO confessed on reading the Yoshida Hearings, "I had the same impression as the videoconference. There were places where I was so disappointed I wanted to cry."

"They knew they were fighting a losing battle, but I got the impression that they were brave people who would continue to make assaults without regard for their own lives. I can only bow my head to them. But ... after all this I think it was down to advance preparations. You have to educate the right people, train them and prepare them. You have to be aware that business as usual and emergencies are different. And it's not one or two, but a team that needs to be prepared. Without this, there's nothing the field can do."

Whether "just the right things here" can be carried off or not will decide whether full use of resources and governance can be made to overcome a crisis. The "logistics issue" is not just one for the materials team. It is none other than a strategic issue.

IV ICS (Incident Command System)

> *A: To put it extremely, venting is just a matter of opening a valve, so I thought we could do it if we could open the valves, but then various things start coming out like there's no air in the AO valve (air operated valve) or the MO valve (electric drive valve) is useless. When you ask if it can be done manually, you're told they can't get in because of the high radiation (...) That's when you finally realize how serious things are. But that doesn't get across when you contact Head Office or Tokyo, the difficulty involved in the venting, so they just tell you to hurry up, to get on with it quickly. There was really a huge gap between the actual site, the Central Control Room site, the quasi-site of the Emergency Response Center and the people far away from the site in Head Office."*
>
> *"The furthest away was the Kantei. They thought they would be opened as soon as a ministerial directive was issued, but that wasn't on."*
>
> (July 22)

> "First of all, why was it the Kantei? Why was the Kantei coming directly here? What was the center at Head Office doing? (...) I thought it was strange all along."
>
> (August 8)

Prime Minister Naoto Kan, Chief Cabinet Secretary Yukio Edano, and Minister for Economy, Trade and Industry Banri Kaieda, who were all ensconced at the Kantei, were frustrated with the TEPCO side, who having announced after 3 a.m. on the 12th that they were going to vent, were taking forever to go into action.

Kan decided he himself would go to the site.
Kaieda decided to have the venting implemented by a ministerial directive.
Yoshida was puzzled by the news of a prime ministerial visit to the site.

"I learnt that Prime Minister Kan was coming when we were in a mad flap about whether we were going to vent or not, and I thought there was nothing we could do about it and asked for preparations to be made (...) Despite our efforts at the site, we were having trouble getting it done, but even if we were told to hold off because Prime Minister Kan was coming, since the site was in a situation where we couldn't get the valves open anyway, we were in the middle of attempting to do so, I thought it would still be difficult for a while (August 9)."

To begin with, confusion reigned in the early hours of the 12th about which unit should be vented first. Preparations were initially made to vent Unit 2. Blueprints of Unit 2 were brought out to inspect where the valves were. However, neither the large valve nor the small valve had handles. As such, it was impossible to open them manually.

While this was going on, they received the information that the RCIC in Unit 2 was operating. They then decided to start venting from Unit 1. Looking at the Unit 1 blueprints, the small valve had a handle. However, opening it manually was not "anticipated".

Yoshida especially couldn't stomach the fact that Head Office seemed to have doubts that the site was hesitating about venting.

"I was as keen as anybody to get water in quickly. That's what I was thinking. But there were procedures (…) I was thinking I'd like to kick the hell out of the guys saying the site was hesitating (July 29)."

On the other hand, he had the following to say about Kaieda's directive.

"To put it extremely, my state of mind was one of you 'do it yourself if all it takes is a directive to implement it' and things at the site weren't going well"

"I don't know whether it was their anger with TEPCO that led to the directive to implement, that's between Head Office and the Kantei, all I can say is I didn't know (July 22)."

The valves were supposed to be opened electrically, but since there was no power those motors could not be used. A compressor was needed to open them by air, but they didn't have one. Therefore, the only option was to open them manually, which meant sending workers into an area of high radiation.

Neither the Kantei nor Head Office knew just how daunting a task the venting process was. In fact, not even the Roundtable in the Emergency Response Center had any idea of what the Central Control Room was going through.

And yet, issuing instructions made both the government and Head Office feel as if they were doing something. This is what triggered Yoshida's genuine anger.

The underlying issue here is the question of governance in crisis management.

If you look at the seating order on the head office side for the videoconferences, the chairman, CEO, executive for general affairs and executive for planning were enthroned in the very middle. The executive for recovery, who in terms of incident command systems should be in the center, was sitting in the furthermost seat. In the U.S. incident command system for nuclear power plant operators, it is the person in charge of recovery who sits in the middle.

However, in the Fukushima nuclear accident, TEPCO CEO Masataka Shimizu occasionally butted in issuing instructions, at times intimidating people with a "this is an order from the CEO". This seems to have been based on an assessment during the Fukushima nuclear accident that the CEO should spearhead Head Office efforts because the Nuclear Disaster Special Measures Law stipulates that the prime minister is the chief executive for the government and the CEO for the operator respectively.

However, a manager with no technical background whatsoever may well be involved in management and overall crisis management events, but it is highly irregular that he should meddle in individual accident responses.

To cite an example, just because the CEO happens to be aboard one of the company's airplanes that is about to crash, it's highly unlikely that he/she is going to grab the control stick.

Moreover, CEO Shimizu was a manager totally unsuited to dealing with a crisis.

To begin with, on the day of March 11, he was away on a sightseeing trip with his wife in the Kansai area. On the evening of that day, the Nagoya-Tokyo bullet train service was at a standstill. Trying to get back to Tokyo from Nagoya on a Self-Defense Forces flight, he requested emergency transport to the Ministry of Defense, but Defense Minister Toshimi Kitazawa refused him. It was after 9 o'clock on the morning of the 12th that Shimizu took up his seat in the Operations Room at Head Office. He had made it to the control tower but bearing a sense of psychological inferiority.

TEPCO failed to establish an Integrated Command System, or "vertical governance", including its relationship with the Kantei, until the very last.

V What is needed is a Command Tower, not a "Study Session"

> A: I got a phone call from Takekuro (TEPCO Fellow) at the Kantei (...) in short saying that the Kantei hadn't yet approved the seawater injection. So, his instruction was to stop the seawater injection. All I heard was stop it.
>
> I asked if Head Office knew about it and spoke with Takahashi (Fellow), and deemed it was unavoidable, so I thought we'll stop, but regarding what stance to take on having already injected it, I consulted about it being a trial run and saying we were confirming if the line was alive or not (...) so the stance was that we had started a trial run at 19:04.
>
> Since I couldn't follow a directive that didn't offer any collateral of when we could resume, I decided to act on my own judgment. So, although I told the fellows at the Roundtable that we would stop, I instructed the disaster prevention group leader who was in charge that I would give the command to stop (...) here, but he was absolutely not to stop and I reported to Head Office that we had stopped.
>
> (July 29)

This is what became famous as Yoshida's "kabuki play" when it was reported even overseas. Let me introduce here an excerpt from my book, Countdown Meltdown (Bungeishunju, 2012) describing the scene.

Around 7 p.m., Takekuro placed a direct call to Yoshida.

Takekuro jumped right in.

"Hey, what's happening with the seawater?"
"We're doing it."

Takekuro was flabbergasted.

"What? Are you already doing it? Stop it!"
"Why?"

Yoshida had already ordered the seawater injection. He couldn't very well put the water they'd already pumped out back into the hose. Takekuro was mad when Yoshida insisted that they couldn't stop it now.

"Shut up and listen. The Kantei's vacillating!"
"What on earth are you saying?"

Yoshida hung up thinking it was pointless listening any more.

Yoshida had been appealing in the videoconferences for the need to inject seawater as soon as possible. Head Office, however, was cautious.

Shimizu rang Yoshida and requested he halt the pumping operation. Yoshida countered.

"We've already started, you know. Didn't I send a fax at 4 o'clock?"

Shimizu replied,

"You can't do it yet. We don't have government approval. You'll just have to stop until we get it."

> "Since it's the wish of the Kantei, please put a stop to the seawater injection. I know there are lots of different opinions, but this is an order from the CEO."

Yoshida was being asked directly by his chief executive officer. He responded obediently, "I understand."

Yoshida then declared,

> "The Kantei has commented on the seawater injection. We're going to suspend it temporarily."

> In the images of the TEPCO videoconference Yoshida could be seen leaving his seat, walking around the ERC and discussing something with the manager in charge of the pumping operation. The officer sat with his back to the Head Office side. Still on his feet, Yoshida could be seen whispering something in his ear.

The essence of this crisis lies in the ambiguity of the decision-making process for crisis response.

Regarding the lack of transparency in the decision-making process for the seawater injection, the then TEPCO Vice-President Sakae Muto, who was at the Offsite Center, confided the confusion to the Government Investigation. Yoshida answered in the hearings that this was understandable.

"Muto apparently thought injecting seawater would be okay, but no-one was sure who had actually decided this (…) Since Muto was at the Offsite Center and didn't know the Kantei's intentions, I thought he made a comment along the lines of is this in consultation with the Kantei, but the situation was that no-one knew who the decision-makers were (August 9)."

Following the site visit by Prime Minister Naoto Kan, the Kantei embarked on a course of micro-management meddling, in evenindividual accident response measures. There is no doubt that this was a huge problem in terms of ICS.

However, the risk of meltdown was increasing and the need to evacuate residents emerged. Nevertheless, TEPCO Head Office lacked the capacity to act as an involved party. You couldn't blame the Kantei for wanting to bypass Head Office at such a time and contact the site directly.

A senior official at the Ministry of Economy, Trade and Industry heading the crisis management at the Kantei at the time looks back as follows.

"The site superintendent was the most desperate about the power station's problems, but when it comes to the evacuation of residents, well then it's the Administration that becomes the most desperate. Both of them were in a desperate situation (...) An exchange of information between the two was essential in order to meet both functions simultaneously. I think we needed to consider more about how to create that mechanism. The reason Kan (Naoto)-san got involved was he was trying to get information. Hosono (Goshi)-san rang Yoshida-san because he too was trying to find out."

Yoshida's Kabuki Play, however, is fraught with problems from the point of view of ICS. Yoshida gave the following response to the question, "[the seawater injection] was a world first, wasn't it?"

"When you reach this zone, since there's no manual, in extreme terms, I thought it was a judgement call, although it would be strange to call it my intuition (July 29)."

This approach was close to Prime Minister Kan Naoto's thinking that there was no theory to crisis response (This remark of Kan's was made in a Q&A session when speaking of the lessons learned from the Fukushima nuclear power plant accident at the Davos meeting of January 2012).

Certainly, manuals and theories may serve no purpose when thrust into a crisis, and Yoshida's assessment at the time may be perceived as "permissible discretion during a crisis".

The seawater injection was conveyed to TEPCO Fellow Ichiro Takekuro, holed up at the Kantei, by METI Minister Banri Kaeda at 5:55 p.m. on that day as a directive based on Paragraph 3, Article 64 of the Nuclear Reactor Regulation Law. However, immediately thereafter, Prime Minister Naoto Kan called a meeting with Kaieda and others in the Prime Minister's office, asking them to look into "whether there was a possibility of re-criticality or not", deciding to gather again in 2 hours' time to draw a conclusion. Takekuro who was at the Kantei construed this as "not having obtained the Prime Minister's approval", and asked Head Office and Yoshida to stop the seawater injection. This only added to the turmoil.

Therefore, Yoshida could claim that all he did was follow Kaieda's instructions (although NISA only documented Kaieda's directive after 8 p.m., an hour or so after Yoshida's Kabuki Play).

As such, it cannot be categorically criticized as an "arbitrary action" on Yoshida's part. On the other hand, the risks involved in such a Kabuki Play should also be borne in mind.

The scope of responsibility of the site superintendent in the field and the CEO at head office naturally differ. Should a site superintendent go against the orders of the head office CEO, and if the result is right, it can be dispensed with as an "heroic exploit", but if, to the contrary, it aggravates the crisis, or creates a secondary disaster, this raises implications and results that exceed the scope of the site superintendent's responsibility.

The failure to clarify the respective authority and responsibility of the departments involved in the crisis response complicated the accident response and halved their efficacy.

When, on the evening of the 14th, Unit 2 reached a critical situation, Haruki Madarame, Chairman of the Nuclear Safety Commission, suddenly placed a phone call to Yoshida.

> Q: At the time of Unit 2. That was part of that, Madarame-san was putting a strong case for doing such-and-such; in the midst of that you received a phone call, which from your perspective must have been an unnecessary annoyance; the instructions have been decided at the Kantei from the Prime Minister down; you're to go with this; it wasn't like that, was it?
>
> **A: That was just a study session.**
>
> Q: It wasn't especially a command tower?
>
> **A: But I wonder what makes this country move. It's a strange country, isn't it?**
>
> (November 6)

As long as the Kantei was intervening in the accident response, the site assumed its intervention was premised on decision-making. However, Yoshida stated with irony that the Kantei was not acting as a command tower, but was merely conducting "study sessions".

"I wonder what makes this country move. It's a strange country, isn't it?"

VI   The Failure of Safety Regulations Governance
(Videoconference footage)

> Q: May I? 1F-san? We've just got some instructions from NISA and since NISA thinks (...) there's a possibility of an explosion in 1F-1, they've sent instructions for you to

> *consider, for example, measures like opening the blowout panel and so on.*
>
> *A: There was nothing we could do. If I was to speak plainly, yes. That time makes me sick to my stomach even now, "All they're doing is issuing rotten instructions like this". That's all they did. Simply mouthing do this or that. Give me a break!*
>
> *Q: Break the rupture disk and so on.*
>
> *A: Whenever NISA came onto the scene, they only said things that made you want to throw up.*
>
> *Q: Nothing direct?*
>
> *A: Nothing.*
>
> *Q: Representatives from NISA, NISA officers?*
>
> *A: No-one was there by then. Not a single person.*
>
> (August 9)

There was a wealth of emotions imbued in the words, "No-one was there by then. Not a single person."

When the accident took place there were eight Safety Inspectors on site. Four went to the Offsite Center and four stayed behind in the Anti-Seismic Building. After the explosion in Unit 1, however, the remaining four also fled to the Offsite Center.

It was METI Minister Kaieda who ordered them to return to the site. While they did return upon receiving these instructions, they fled once again to the Offsite Center when Unit 3 became critical.

"First of all, there was a phone call from Muto where he said NISA was coming and they may have come once for a short time. Sometime around the 14th. I remember that. They were there briefly. When the Offsite Center was to be moved to Fukushima, everyone pulled out to Fukushima and eventually by the 16th until about the 17th, when the Self-Defense Forces and the fire brigade were rushing in, I think they were all gone (August 8)."

The responsible government officers were in the field "briefly", and then disappeared like the wind. Yoshida criticized NISA using the most insulting of terms that "they were dirty". This was directly triggered by their diffident response to seismic standards.

"All they did was set up a seismic evaluation subcommittee and the like, line up some professors, get the power utilities to draw up some papers, make a report, knit pick, say do something about the bits where comments were made, clear this up and so on. In short, NISA itself did absolutely nothing like determining the criteria itself. Those people never take responsibility, you know (August 8)."

NISA was broken up after the accident, many of its staff moving to the Nuclear Regulatory Agency. Since it was abolished as an organization, NISA has not investigated the pros and cons of its response at the time of the accident.

The same can be said for the Nuclear Safety Commission, which alongside NISA was another of the nuclear regulatory agencies.

Yoshida was unforgiving towards the Nuclear Safety Commission as well. The brunt of Yoshida's attack was directed at Chairman Haruki Madarame.

Yoshida had the following to say about the phone call he suddenly received from Madarame on the evening of March 14.

"He was in a panic (...) I thought who is this bloke, but realized that apparently it was Chairman Madarame, so I was listening along (...) he's saying there's no leeway, you've got to get the water in fast, get it in. When I told him we were currently involved in the venting operation, he said there's no time to vent, get the water in fast! (August 9)"

Part 1   A Reading of the Yoshida Hearings

Madarame was urging Yoshida to get the SR (steam regulator) valve open, depressurize and inject water as soon as possible. However, the officers in charge at the site had pointed out that because the water temperature in the suppression chamber was over 100 degrees, it was likely that the steam would not condense and they would be unable to depressurize sufficiently.

On the other hand, Yoshida had received a report that although they were also trying to vent, the vent valve would not open. It was at this juncture that Shimizu interrupted, and issued his CEO order to "follow Chairman Madarame's method."

While Yoshida described Madarame as "in a panic", one of the team leaders at the Roundtable who heard Madarame's remarks told me that was not his impression. Even if Madarame was "in a panic", he was probably not alone. When the report was received that the SR valve would not open, silence descended on the nearly 200 people in the Emergency Response Center. The sense of do-or-die that struck the Roundtable at that time can be learnt from none other than the Yoshida Hearings.

Moreover, Madarame himself was isolated without any support.

According to Madarame, after the accident broke out, he was hurriedly called to the Kantei, but "there was no-one to consult with, no handbooks or reference materials. I was placed in a situation where I had to assess things totally unaided (Independent Investigation interview)."

Just as was the case with NISA, the Nuclear Safety Commission also failed to function properly in the crisis. The root of this was its unwavering belief in the "safety myth".

The principal responsibility for safety lay with the power utility operators and ensuring this had used a division of duties where it was the job of the NISA to oversee the utilities and that of the Nuclear Safety Commission to establish safety regulation guidelines. This system itself was premised on safety regulation in normal times and neither NISA nor the Nuclear Safety Commission could function adequately in an emergency such as a severe accident.

The occurrence of an emergency would lead to the admission that prior accident prevention was insufficient.

Severe accident response had gone virtually without review since 1992 when the Agency for Natural Resources and Energy requested the voluntary efforts of the power companies in response to the decision of the Nuclear Safety Commission. The accident management procedures drawn up by the Nuclear Safety Commission only focussed on internal events such as a mechanical failure or incorrect operation, and did not take into account external events such as tsunami, earthquakes, or terrorism. This, therefore, allowed NISA and the power companies to treat external events as "unanticipated".

To begin with, since the guidelines drawn up by the Nuclear Safety Commission are not legally binding and are nothing more than "guidelines", the NSC has neither the means nor the measures to enforce them.

The complaisance at a time of crisis of NISA and the Nuclear Safety Commission – to which must be added the Ministry for Education, Culture, Sports, Science - was made painfully obvious to Yoshida and others in the field.

All they could issue during a crisis were "rotten instructions". This vividly portrays the failure of nuclear safety regulation governance.

VII Who will "put their life on the line"?

> *Q: So, the explosion took place just as the induction was over and you were about to try getting water into the back-wash valve pit?*
>
> *A: Yes. First, we heard from the field that forty some people were missing. That was the first report immediately after the explosion, and I wanted to die at the time. If it was true that forty some people had died, I was thinking of committing hara-kiri. Gradually we got more information, and the number of missing got closer to zero. Some staff from the manufacturers had been injured, four or five of ours and the associate companies, in short, no one lost their life.*
>
> *Even with debris flying everywhere, no one died at the site. I could only think it was thanks to Buddha.*
>
> (July 29)

March 14, 11:01. The Unit 3 building exploded.

A huge amount of concrete fragments was scattered all over the place after the explosion, wounding some of the workers working near Unit 3. More than 40 people were missing...

This was the first report to come in from the field to the Roundtable. An engineer at the Roundtable at the time described the situation then as follows.

"We learnt of the explosion at Unit 3 via television footage. Unlike Unit 1, it was huge. After some 20 minutes workers who had been outside started coming back in. Some were bleeding from their forehead, others broke down in tears; they were from the in-house fire brigade, the maintenance team, the civil engineering team. As people started saying I'm so-and-so, my name is such-and-such, I felt a real sense of relief that there weren't 40 people missing."

If, at this time, there had been fatalities, the subsequent crisis response at Fukushima Daiichi would have taken on a completely different aspect. One of Yoshida's "lieutenants" remembers hearing Yoshida murmur, "we probably won't be able to continue emergency accident response if we have fatalities."

Members of the Self-Defense Forces, who had arrived in the vicinity of Unit 3 in order to supply water, were injured at this time. If, however, there had been SDF fatalities at this time, then any subsequent mobilization of the SDF would have been much more difficult.

After the explosion in the Unit 3 building, all operations were stopped to confirm no fatalities, and after sending the injured to J-Village for treatment, Yoshida spoke to everyone packed into the Emergency Response Center.

Yoshida's statement at the hearings:

"Everyone was stunned, I mean it was like their mental processes had stopped. So, I got everyone together and told them to resume work in such a state; my judgement was wrong; I'm sorry, but the water injection is probably at a standstill at the moment; we have to prepare for the water injection at Unit 2; things will only become worse if we leave it unattended; so, I bowed my head to them and asked them to go back out to the site (…) and take firm measurements of the radiation, to remove the debris, to change the minimum number of hoses needed for the water injection and so on. And what really moved me was that upon saying this, everyone made a move to go out into the field."

The aforementioned engineer says he can't forget Yoshida

standing erect and appealing to everyone, "I'm sorry to put you in danger, but Unit 2 is in a critical state. Can you muster up the courage to go back there again?"

"He was begging them."

They nodded and left the room in silence.

At the time of the accident, Site Superintendent Tadayuki Yokomura at the Kashiwazaki-Kariwa NPS, who was providing indirect support for the accident response, said there was a specific fear at Fukushima Daiichi due to the design structure of the reactor. For example, the donut-shaped suppression chamber attached to the bottom of the containment vessel did not have a shield. And because the containment vessel was exposed with just a thin iron plate, the heat came directly through to the outside, as would radiation.

He said, "A normal human being would be too afraid to go near it. And yet the death squad went despite not knowing what was happening to the core in that state. That was the 1F spirit."

Site Superintendent Naohiro Masuda, who was spearheading efforts at Fukushima Daini, also stated that to issue the order to "go to the site" was "the most difficult decision I faced." "You are sending your subordinates out into the dark where goodness knows what was lying about in a situation where there were still aftershocks and tsunami warnings. I later heard that the devastation at the scene was unimaginable, so much so that I was told 'How could you have sent them in there?'" (Funabashi, Nuclear Defeat - Crisis Leadership, Bunshun Shinsho, 2014)

Although they learnt that "40 people missing" was the wrong information, two employees working nearby were injured by the explosion at Unit 3. This news severely upset the operators at the Central Control Room among others.

A TEPCO executives recounted it as follows.

"We're not soldiers, so we've never experienced the guy shooting next to us falling down dead, so a situation like that was upsetting. The kind of training we probably need in future accident training is how to stay calm and continue responding to the accident even if you have lost a colleague."

Who is willing to lay their life on the line and keep working during a crisis?
Who can issue instructions like that and on what grounds?
What should you do if you don't have grounds like that?
Japan had never faced head on an ultimate challenge like this. There is no security system in place on either a national or societal level.

Japan's national efforts to anticipate threats to the safety (security) of all life and safety, and to protect the whole from such threats is insufficient.

One of the reasons the Japanese government refused to introduce B.5.b. measures like the US government in the face of nuclear power plant terrorism after the terrorist attack of 9.11 was said to be concern over the vested interests of the "economic agency" NISA and bureaucratic culture that was "scared" by B.5.b, which was put in place as essential security (a senior nuclear energy policy official at the Ministry of Economy, Trade and Industry).

In short, they avoided thinking about that as being "unanticipated" for the safety of nuclear power plants. In the same way, even the security of the state does not constitute a body of national security.

It is not that there is a problem with the Constitution.

Articles 9 and 13 of the Constitution clearly state that the protection of the people's lives and safety is the greatest duty of government. Article 13 of the Constitution may well be said to be a provision on national crisis management.

The problem is that there is no national consensus about the concept and state of security to ensure its working. The Fukushima nuclear accident put this kind of ultimate but long-avoided issue fair and square before Japanese society.

A TEPCO executive confided hesitantly to me, "The Fukushima accident taught us that, in the end, we could only rely on ourselves. That's our resolve as we make a fresh start," but no such resolve has been heard from the government yet.

VIII A Severe Nuclear Accident is Always a National Crisis

A: Unit 2 was going to melt, completely breaking the pressure in the containment vessel and all of the fuel leak out if we didn't get water into it. If that happened, all of that radioactivity would be scattered around making the worst accident. Not in the class of Chernobyl, and although it wouldn't be a China Syndrome, that would be the situation. And then, we'd have to stop putting water into Units 1 and 3 (...) Eventually, we'd have to withdraw from here (...) Mentioning that in the Anti-Seismic Building would have struck fear into everyone, so I think I told Muto over the phone.

I thought Unit 2 had had it. At this point, to be honest (...) I mean, we couldn't get the water in. And if we couldn't get the water in, well, that meant the fuel would melt. Fuel melts when it reaches 1200 degrees (...) The walls of the pressure vessel break down and so do the walls of the containment vessel here and there (...) All of the fuel escapes. Whether it's plutonium or whatever, it'd be worse than the cesium mentioned now. All of the radioactivity would escape and be scattered all over the place, so "our

> *image was the destruction of Eastern Japan".*
>
> (August 9)

TEPCO CEO, Masataka Shimizu, first mentioned a "worst case scenario" around 5 p.m. on March 14. He said in a videoconference,

"Well, a worst-case scenario for Unit 2. You need to draw this up and response measures, so the maintenance team at the back office get a firm grasp of this and report please."

Shimizu had been startled to learn that the "worst case" would arrive in two hours' time.

About that time, the water level in the Unit 2 pressure vessel had reached the nuclear fuel's TAF (Top of Active Fuel), and had fallen ten millimeters below it by 17:30. The nuclear fuel was finally starting to become exposed. Pressure in the reactor core was also on the increase. No matter how many times they tried, the valve in Unit 2 refused to open.

They only had "about two hours" until it became the "worst situation". Even Yoshida was becoming discouraged.

One of the workers at the Roundtable told me they could still see the figure of Yoshida leaving his seat at the Roundtable and standing in the back in a trance-like state.

Shimizu must have been devastated on hearing "in two hours' time".

It is not known whether Shimizu shared Yoshida's vision of "the destruction of Eastern Japan" at the time or not. Nevertheless, the crisis management at TEPCO were forced thereafter to assume a "worst-case scenario".

It was the same for the Kantei.

The Kantei had also started feeling the need to create a "worst-case scenario" around that time, and had secretly asked the Nuclear Safety Commission Chairman, Shunsuke Kondo, to draw one up.

The document (Sketch of the Fukushima Daiichi Nuclear Power Station Contingency Scenario) the Independent Investigation obtained assumed a case where the fuel pool at Unit 4 would be destroyed, and following core concrete interaction, the same thing would occur in the other fuel pools.

If the situation progressed up to this point, accident response at the Anti-Seismic Building would no longer be possible. The document proposed a final resolution of "shielding with a mixture of sand and water" the reactor in melt down and the fuel pools. It also assumed that, if this scenario became a reality, a radioactive cloud (plume) would reach the Tokyo metropolitan area.

But, who would perform the "shielding with a mixture of sand and water" operation in such a case?

Japan does not have a rapid reaction force in the case of a severe nuclear power plant accident. There was no such function at the regulatory agencies, such as the Nuclear and Industrial Safety Agency.

It was around March 17 or 18, when the police, SDF and fire fighters were squabbling over procedures for the water injection operations at the fuel pool in Unit 3, that Eiji Hiraoka, NISA deputy director-general, was challenged by staff from other agencies around the roundtable for the Emergency Team at the Kantei's Crisis Management Center as to why NISA wouldn't even get rid of the debris.

It was all Hiraoka could manage to say that,

"NISA doesn't have any operational units. NISA officials are administrative officers first and last. They're only inspectors."

The Government had no choice but to leave on-site crisis response up to TEPCO.

The US government was startled by this aspect. It severely questioned "how long did the Japanese government intend leaving things up to TEPCO?"

Mike Mullen, US Chairman of the Joint Chiefs of Staff, told Ambassador Ichiro Fujisaki Ichiro by way of an ultimatum,

"Why, in such a serious crisis, is the Japanese Government leaving the accident response up to TEPCO? Isn't it for times like this that you have troops? Why doesn't the Japanese Government use the Self-Defense Forces? It's incomprehensible. If things keep on like this, we may have to think about bringing the US military home."

Prime Minister Naoto Kan stormed into TEPCO Head Office a bit after 5:30 a.m. on the 15th. And in front of the TEPCO employees lined up in the Operation Room, delivered a "Prime Ministerial Speech" lasting nearly 10 minutes.

If TEPCO withdrew from the Fukushima Daiichi Nuclear Power Station accident response, Japan might be "occupied" by foreign powers, such as the United States and Russia. After mentioning such a fearful scenario, Kan said,

"You are the ones who have to do it. Put your lives on the line."

Kan later described to me his fear at the time.

"[At Chernobyl, in the end] the Soviet Union used the army. In Japan's case, there was absolutely no choice but to use the Self-Defense Forces. If you didn't, somewhere would come in by force. Once that happened, Japan's meaning as a state would cease to exist."

Haruki Madarame, Chairman of the Nuclear Safety Commission, has said that Kan "lost it" at the time, and Kan may well have fallen into a psychological state close to that of elite panic. However, it is not possible to criticize him for overestimating the deterioration of the situation.

Regardless of whether one likes it or not, a severe accident at a nuclear power plant will become a national crisis. It always leads to a national crisis.

As the prime minister of the country, he was forced to consider the "worst case scenario" and to demand of TEPCO, "You are the ones who have to do it. Put your lives on the line." And if that didn't work, there was nothing for it but to give the command to the dispatch the SDF.

In fact, in the face of the gravest national crisis since World War II, Kan turned to the SDF as the "last bastion."

The Self-Defense Forces had already become the "first line of defense" in the field for the quake, the tsunami and the nuclear accident. There were many villages, towns and cities where the tsunami had washed away all trace of town halls. Administrative functions had collapsed in many municipalities. It was only the SDF that could help with those functions.

"At the request of local governments, the Self-Defense Forces transported unidentified bodies and buried them. However, since there was initially no legal basis for this, a high-ranking SDF officer revealed that in some places, they were told by police officers at the scene that they were being 'arrested red-handed'. Regarding unidentified bodies, Article 92 of the Family Registration Act stipulates that, 'a police officer must draw up a written necropsy statement, and attach this to the death report which shall be submitted to the head of the local government in the district where the death occurred without delay', and designates the municipality receiving the report to transport and bury the remains. However, many municipalities felt into dysfunction in the Great East Japan Earthquake, the SDF having no choice but to perform as an emergency measure the handling, transport, and burial. (Later, the Government legalized these kind of emergency measures by the SDF with the promulgation of Special Measures for Burial and Cremation Permits Concerning Cemeteries and Burials Affected by the 2011 Tohoku-Pacific Ocean Earthquake.)

This episode illustrates just how unprepared Japan was in systemic terms for an emergency.

Four years have passed since the accident. And Japan still does not have a rapid reaction force for severe accidents at nuclear power stations.

The Independent Investigation recommended the following in its report (February 2012).

"The responsibility of the state in the case of a severe nuclear accident must be to clearly position in the legal system the role of the execution units responding at the time. It should aim at creating in the future a full-scale emergency response unit for severe disasters and accidents similar to that of the United States' Federal Emergency Management Agency (FEMA)."

Upon passing the bill to establish the Nuclear Regulatory Commission and the Nuclear Regulatory Agency in 2013, the House of Councillors passed an addition resolution that, "Based on the extensive damage caused by the Great Eastern Japan Earthquake, in order to enable a more agile and effective response to large-scale disasters, including a nuclear disaster, the Government is to undertake a thorough review of government institutions responsible for large-scale disaster response including referring to the United States' FEMA, and to implement the necessary measures based on those results", thereby asking for an investigation into the need for emergency troops, but that is all.

The "role and responsibility" of the state in times of emergency is still not clearly defined.

The role of the first responders to a severe nuclear power plant accident also remains ambiguous.

IX The National Policy/ Private Management Trap

A: I never said anything about leaving with an all-out withdrawal. I was staying and, of course, I'd keep the operators, but all I said was please come up with some ideas for assuming the worst, and that I'd evacuate people not involved.

At the time (...) I also heard talk that CEO Shimizu had said to Kan "Let us withdraw". I don't know whether that's linked to something I said to someone at Head Office being passed on to Shimizu or what I said to Hosono-san, but there was some kind of double-line talk.

Speaking about that withdrawal commotion, I thought what nonsense are they carrying on with now. Let me say something and that is, we didn't run, did we? Say it if we'd turned and run. I don't know what sort of Mickey Mouse discussions were taking place at Head Office or the Kantei, but did the site run away? No, it didn't (...) Since the situation was extremely dangerous, we would have had to withdraw at the very last if things got bad, but I would have kept the minimum staff for water injection and so on. It was also my intention to stay.

Q: Can we interpret that as meaning that the response of everyone at Head Office from CEO Shimizu down, the executives was along similar lines?

A: Our side didn't know whether that person [CEO Shimizu] went to the Kantei or not.

(August 9)

On the evening of the 14th, Site Superintendent Yoshida Masao decided to temporarily withdraw workers from Fukushima Daiichi Station.

"I called in staff from personnel affairs, and quietly into a room at that, and told them to check how many people there were. (…) I told them to check particularly the number of people not involved in operations and repairs. Not that we could do anything about the people assigned from Head Office. How many buses could we use? I thought there were two or three, but what about drivers? Were they fuelled up? I told them to have them wait out front. (…) Those were my instructions (August 9)."

The room Yoshida was referring to was the duty room on the ground floor of the Anti-Seismic Building. There were four rooms fitted out with beds, but after the accident, one of them was set up as Yoshida's personal room.

Yoshida has repeatedly asserted that it was his intention that the several tens of people who formed the nucleus of operations and repairs would stay and continue with crisis response. He stated, "we would have had to withdraw at the very last if things got bad, but I would have kept the minimum staff for water injection and so on. It was also my intention to stay."

Yoshida's words are probably true. Two of his most faithful subordinates supporting him in the Emergency Response Center also attested the same to me. One of them said when Prime Minister Kan stormed into TEPCO Head Office in the early hours of the 15th and he heard him via the videoconference yelling at the TEPCO executives lined up "there will be no withdrawal", that he "felt genuinely uncomfortable."

In the first place, at Fukushima Daiichi NPS, the expression "withdrawal" itself was assumed to refer to returning operators to the Anti-Seismic Building. Yoshida divulged during an interview with members of the Diet Investigation Commission that he was surprised that the Kantei had interpreted the word as meaning a "withdrawal" from the Anti-Seismic Building to outside of the plant at Head Office.

In the Government Investigation hearings, Yoshida said of the Kantei and especially Prime Minister Kan, "what nonsense are they carrying on with" and Mickey Mouse discussions at Head Office and the Kantei", expressing his distrust of Head Office in the form of putting it on the same level as the Kantei.

Yoshida has testified as follows.

"At the time (...) I also heard talk that CEO Shimizu had said to Kan "Let us withdraw". I don't know whether that's linked to something I said to someone at Head Office being passed on to Shimizu or what I said to Hosono-san, but there was some kind of double-line talk (August 9)".

Yoshida is unclear on from whom he heard the information that "CEO Shimizu has asked Kan to let us withdraw". It is possible that such information was going around the site about that time.

One of the engineers holed up at the Roundtable has stated, "There was no talk like that near the Roundtable around the time. I reckon Yoshida heard it later." (He also added, "Of course, sensitive subjects like evacuating or withdrawing weren't dealt with by videoconference, but through direct interaction with the parties concerned by Yoshida over the phone. The only time he mentioned withdrawal on the videoconference was after the explosion in Unit 4 (just after 6 a.m. on the 15th).

There are no real facts that "CEO Shimizu asked Kan to let us withdraw."

Shimizu's actions that night, after being told the "worst situation is in two hours' time", as well as the real story of discussions among TEPCO's top management from the evening to the night of the 14th concerning "evacuation/withdrawal" remain shrouded in mystery, but Shimizu placed calls alluding to "withdrawal" first to NISA Direct-General Nobuaki Terasaka, then METI Minister Banri Kaieda, and also to Chief Cabinet Secretary Yukio

Edano. He was to ring Kaieda repeatedly that evening. (However Terasaka told the Independent Investigation Commission "I don't believe [CEO Shimizu] mentioned withdrawal.

However, Shimizu did not telephone Kan directly that evening.

Yoshida often displayed what was close to contempt for Shimizu.

"There was a call from the Kantei and Madarame-san came on the line saying inject that water quickly; get the pressure down and the water in; (…) there wasn't a lot of explanation, just depressurize, inject water and Shimizu just happened to be listening to the videoconference at the time, and started yelling to do what Chairman Madarame said. I thought what a cheek this guy has, he knows nothing about the field (July 29)."

So much so that you get the impression that CEO Shimizu only appears in the Yoshida Hearings to be abused.

After the Kantei, and especially the Prime Minister, started intervening directly in the accident response, the responsible officer from TEPCO for "consultation" with the Kantei (especially the Prime Minister) had to be the CEO. The pattern becomes convey this to the CEO, we'll have to get the CEO to do this, irrespective of the nature of the "consultation". Shimizu, however, was not the kind of leader up to such a responsibility. Those in the field knew this only too well. Yoshida's failure to use the "honorific" was infused with this kind of distrust.

Standing in Shimizu's shoes, however, as top management he was responsible for the lives of his employees, and from this perspective he too had to deal with crisis management.

There was no doubt that they had reached a stage where hard choices needed to be explored about how to act in concert with the government in a "worst-case scenario". He must have been pushed to evaluate how to mitigate the risks as much as possible, bearing in mind too the possibility of any and all litigation risks.

If, in the very end, they were forced to "withdraw", that was not a decision TEPCO alone could take. An out-and-out abandonment was not possible. It would not be strange if Shimizu, who had dealt with the crisis response by shouldering all the government instructions including "do what Chairman Madarame says" for the Unit 1 venting, the seawater injection, and opening the SR valve in Unit 2, and making them orders from the CEO thought to "obey" government directives for "withdrawal", the ultimate decision. It would be even stranger if the Planning Department, that acts as TEPCO's government lobbyist corps, did not think of such a risk hedging measure.

An engineer from one of TEPCO's associate companies, who was responding to the accident alongside TEPCO staff in the Head Office Operations Center at the time, gave his impression of Shimizu's actions as "I think he was looking for an escape, that he had done this based on instructions from the state."

One thing that is certain is that TEPCO Head Office never issued clear instructions to the field on how many people to leave behind from which departments in a "worst-case scenario". That is, they did not show a clear intention to remain in the field until the end as a judgment from Head Office.

However, Edano and Kaieda, who received phone calls from Shimizu, both testified that they didn't understand what Shimizu was trying to say because he was so vague. Shimizu was clearly in a panic and was not in a condition to judge calmly. Yoshida didn't bother dealing with such a Shimizu, and consulted exclusively with Vice CEO Sakae Muto about an evacuation scenario not via videoconference but by mobile phone.

On the night of the 14th, when Muto instructed his subordinates to draw up an evacuation plan, it is said that he tossed the job back to the site saying, "This should be decided by the field." Yoshida, however, was flat out handing the accident response and didn't create a formal evacuation plan for each individual because he didn't have the time to put it down on paper.

Yoshida was prepared to stay in the Anti-Seismic Building, confessing to the Diet Investigation Commission that interviewed him in hospital that "in the very end, I thought there were probably about ten people who would die with me."

A cardinal rule of crisis management is to prepare without being over specific and to promote mutual understanding to the point of repetition.

On the other hand, it cannot be denied that the poor communication akin to a telephone game inside TEPCO, which irritated the Kantei intensely, also applied to the Kantei itself.

Although METI Minister Kaieda Banri, who received a call from Shimizu, initially took Shimizu's remarks as meaning "a temporary withdrawal from Daiichi to Daini", he trembled at a scenario where "after the temporary" would come "a complete withdrawal", and conveyed this feeling to Prime Minister Kan. Edano also assumed Shimizu's call was not just about a temporary withdrawal but was seeking approval for a more full-fledged withdrawal, reporting to that effect to Kan. Edano replied at a Government Investigation hearing, "I'm quite confident that it was definitely about a full withdrawal... I mean, there was no special need to pass on to the cabinet secretary that other non-necessary personnel would flee. There's no possibility of a misunderstanding." (Asahi Shimbun, An Outline of Politicians' Survey, morning edition, September 12, 2014)

Both Kaieda and Edano have testified that Yoshida's determination to stay in the field "didn't reach" the Kantei.

Even so, what is strange is why Edano didn't thoroughly scrutinize the information about a withdrawal. If Shimizu's remarks were ambiguous, why didn't he check with TEPCO Chairman Tsunehisa Katsumata? Why didn't Hosono take on this task? Hosono was already keeping in touch with Yoshida by phone.

Nuclear power generation in Japan has been promoted under the banner of "National Policy/Private Management" in a framework where the "national policy" of promoting peaceful use of nuclear energy championed by the government was carried out by private companies operating nuclear power generation businesses under "private management".

While "National Policy/Private Management" may have been fine for normal times, it completely failed to work in an emergency. The nation revealed its lack of ability to manage a crisis. Is "National Policy/Private Management" nothing but a system that obfuscates responsibility in a time of crisis? This has led the people of Japan to question whether it is safe to entrust nuclear administration to such a government.

The Government is yet to answer this fundamental question.

x   What is a First Responder?

A: First we had the riot police come, but they were not very useful. And they only came once, then they were pulled out. It was only after a lot of fuss that they came and then they were pulled out; in short, they didn't have an impact.

Q: How about the Self-Defense Forces?

A: To be frank, speaking in hindsight, it was all pointless. Even if they could pump all their water in, we were talking of volumes in the order of 10 or 20 tons, and when you take into account the surface area of the fuel pools, it was pointless even if they got it all in.

This was especially true of the Fire Department equipment, and although it started off all right the end of the

> *hose would gradually start drooping. But even when you told them it was falling, they didn't go to fix it.*
>
> Q: *That was the Fire Department, was it?*
>
> A: *The Fire and Disaster Management Agency. I'll be frank. It was the heroic Fire Department.*
>
> (August 8, 9)

As Units 1, 3 and 2 in turn went into meltdown, the focus of the crisis management shifted to the spent fuel pools at each unit.

A large number of fuel rods were laid out in the fuel pools. Just as with the reactor cores, they needed to be cooled. In the early hours of the 14th, the temperature of the spent fuel pool at Unit 4 had risen to 84 degrees. Deeming that the site could "not handle" the spent fuel pools, Yoshida requested, "Head Office come up with something, get the SDF to help, anything."

There was a strong fear that if it boiled dry, the fuel would melt out, react with the concrete, and spread radioactive material outside.

Sometime after 6 a.m. on the 15th, a booming sound rang out at the Fukushima Daiichi site. Yoshida as testified in the hearings that on the spur of the moment he thought, "I wonder if the fuel in the spent fuel pool has overheated and broken out."

(When a report was received in the early hours of the 12th that something like a white haze had been sighted inside the Unit 4 building, some of the engineers encamped at the Roundtable intuitively felt "something's wrong with the Unit 4 pool", and started considering measures for injecting water into the pool.)

Masaya Yasui, who had decamped from METI to the Kantei and was serving as a technical advisor for Kan and the other political staff at the Kantei, was more afraid than anything else at the time that "If there was a meltdown in the fuel pool, even the emergency measures being carried out at Units 1, 2, and 3 would become untenable. That was the greatest crisis. If that happened, there'd be no-one left in the area."

Yutaka Kukita, acting chair of the NSC, pictured a scenario where "the nuclear fuel melted, and then a fire would break out destroying the bottom of the pool from where fuel would splash out, that would be the worst."

Of all the fuel pools, the one at Unit 4 was the subject of especial attention, holding as many as 1,331 spent fuel rods, which had only recently been removed. In particular, it was the possibility of the fuel pool at Unit 4 boiling dry that the US government considered the "worst-case scenario". On the night of the 16th, even after the Japanese government and TEPCO officers decided "it looks like there's water" upon seeing the aerial images provided by the SDF, the US continued to guard against the risk of the Unit 4 fuel pool.

However, given that there seemed to be water in the Unit 4 fuel pool, the Japanese government decided that it was more urgent is get water into the Unit 3 fuel pool, and hurried the water injection there.

It was the Integrated Response Office (headed by Prime Minister Kan Naoto) set up at TEPCO Head Office on the 15th by the Government and TEPCO that acted as the command post for the water injection strategy. It was the first responders of the police, the Self-Defense Forces and firefighters who carried out the pumping operations.

Of the first responders, time was wasted between the police, the Self-Defense Forces and the firefighters about who was to do what and when in the fuel pool operation. By rights, the pumping operations should been concentrated on the professional firefighters who were experts at this from the start, but their turn came last.

Hearing that the SDF was being dispatched for the water operation, Yoshida and his on-site team in the Emergency Response Center were grateful, "that'll be a help."

However, time was lost due to the fact that "the SDF and the riot police took their time getting to Fukushima Daiichi." Yoshida looked back at the hearings, "clearly, nobody wanted to come to a high radiation area. Especially the Fire Department."

The Fire Department mentioned here refers to the Tokyo Fire Department's Hyper Rescue Squad. They arrived on the scene in the evening of the 18th, and began the water operation late that night.

Apparently, Yoshida had high expectations for the pumping operations of the Hyper Rescue Squad as well. He had a video camera shoot the water injection operation, and applauded when the first water discharge landed in the pool. The other senior officers at the Roundtable followed suit and clapped.

However, the nozzle of the high-pressure pump soon dropped, and disappointment spread as it become apparent to everyone that not much water was getting into the pool. Moreover, even though the situation had not improved at all, the Hyper Rescue Squad returned to Tokyo on the 19th, and its captain gave a press conference. Helped by the large tears the captain shed at the time, the Hyper Rescue Squad were suddenly treated as heroes. There was a viciousness to Yoshida's remark about "the heroic Fire Department".

In short, in the words of Yoshida, the pumping operations of the police, the Self-Defense Forces and the firefighters were nothing more than a "haphazard strategy" in terms of continuously injecting a large volume of water into the pool.

However, if one was to defend the first responders, they plunged into an unknown world of radiation and carried out the pumping operation. The crucial point for first responders is that it is only by protecting their own lives and those of their fellow workers that they can save the lives of people. Bearing in mind the risk of exposure, it was only natural for them to exercise the utmost caution. It is not incomprehensible that starting with Yoshida, people at the site were critical of the first responders given the fact that they themselves were badly exposed waiting outside the building to guide them through the site as well as their frustration with the "haphazard strategy". However, the relief power-line distribution workforce rushed in from TEPCO Head Office also strongly insisted they would not be able to carry out their work from the point of view of exposure risk management if they weren't provided with guides from the site personnel. In order to take full advantage of external support, on-site guidance and direction is also the site's responsibility.

It was after the arrival of concrete pump vehicles from the German company Putz Meister that continuous water injection into the pool became possible. With a 58-meter long arm, they were able to pour water in pinpoint from above. Because the arm looked like a long neck, they were referred to as giraffes.

From March 22 on, starting with Unit 4, then Units 3 and 1, being able to use these pumps made it possible to contain the fuel pool crisis just barely.

One of the things that became apparent in the crisis management of the Fukushima nuclear accident was that the issue of command authority among first responders was the most difficult area of all the coordination of interests between bureaucratic institutions.

Even in the midst of such a national crisis, the government was unable to solve this problem, the fuel pool water injection strategy taking on the hue of a performance for each of them to "make a show" of "acting in concert".

The results of the post-verification by the police are self-congratulatory about the police's water injection operations "leading the way" for those of the SDF, the Tokyo Fire Department and so on. However, while the Independent Investigation report does to a certain extent recognize the water injection operations of the first responders, it also raised doubts regarding the water injection operations of the police, "in a field requiring large volumes of water discharge in an elevated place, was there really any need to pump water using equipment for a completely different purpose?" It also pointed out regarding the water injection operations of the firefighters, "doubts remain over the fact that organizing forces around the firefighters, who had vehicles capable of continuous water discharge, was not considered."

Neither the Government Investigation nor the Diet Investigation has conducted an evaluation of the first responders' crisis response.

The police, the Self-Defense Forces, and the firefighters all continue to be a "blank area of investigation" in the Fukushima nuclear power station crisis.

XI "Mental Paralysis" Creates Rigidity in Behavior and Ideas

> *A: Since "15 meters is a level that causes mental paralysis", and while lots of people are talking about AM for this or how to respond to that now, and since I get beat up everywhere these days if I mention the term "unanticipated" so it's hard to use it, it's not something you can just dismiss as an operational thing, so you fall into mental paralysis.*
>
> *Depending on the size of the tsunami, covering for it, there are some covers that are just meaningless. After all, you consider various things, but, ultimately as long as you're not given a design base that anticipates this kind of tsunami, well, then, a true deliberation just isn't possible."*
>
> (November 6)

A huge tsunami flowed into the basement of the building during the Fukushima Daiichi accident, flooding the switchboard and causing a total loss of the AC power supply. Had preparations for a tsunami been in place, the core meltdown could probably have been prevented.

In fact, despite the fact that the Tohoku Electric Power Onagawa Nuclear Power Station was located on the same Pacific Ocean coast, and was hit by a tsunami more than 12 meters high, because the site was above sea level, its buildings were spared from being swallowed up by seawater. The Onogawa NPS was saved because, in preparation for a tsunami attack, the plant had been built on high ground.

Since the Tokai Daini Nuclear Power Station (The Japan Atomic Power Company) in Ibaraki Prefecture had, at the request of the Prefecture, raised its tide wall in preparation for a tsunami, its equipment escaped flooding at the last minute, separating its fate from that of TEPCO (It needs to be investigated why Ibaraki Prefecture could do this but not Fukushima Prefecture).

TEPCO had been warned repeatedly by experts of the high risk of an onslaught by a tsunami exceeding anticipated proportions. In July 2002, the government's Headquarters for Earthquake Research Promotion announced its view that a tsunami earthquake might occur in the offshore trench stretching from Northen Sanriku to the Boso Peninsula, including offshore from Fukushima Prefecture. Research results on the huge tsunami that hit the Tohoku coast in the Jogan Earthquake of 869 were also clear.

TEPCO's in-house experts also started calling for new tsunami measures. Their grounds were, if an earthquake of similar proportions to the 8.2 magnitude Sanriku earthquake and tsunami that struck the Tohoku region in 1896 occurred along the trench off Fukushima Prefecture, the maximum height of the tsunami that would reach the Fukushima Daiichi NPS was estimated at 15.7 meters.

In response to this, an in-house review meeting was held around

June 2008 to discuss this issue. The Nuclear Facilities Management Director at TEPCO Head Office at the time was Masao Yoshida.

The Facilities Management Director was in charge of equipment management, repairs and accident response measures. In the in-house review, Yoshida showed reluctance saying that building a seawall could result in a sacrifice for the surrounding villages and was socially unacceptable.

March 11, 2011. A tsunami of exactly 15 meter's height came rushing in.

Yoshida's response regarding the lack of tsunami measures in the hearings was terribly confused.

First, he confessed to falling into a state of "mental paralysis" because a scale of 15 meters was "unanticipated" and exceeded the level of operational response. On this point, Yoshida was frank.

However, 15 meters was a level that experts had pointed out and was not "unanticipated". As an ex-facility management director, it was unimaginable that Yoshida could fall into a state of "mental paralysis".

Secondly, he tried to justify the then postponement of a decision (non-decision) by saying, "as long as you're not given a design base that anticipates this kind of tsunami, well, then, a true deliberation just isn't possible."

During the hearing he was questioned about "why, for example, didn't the architectural earthquake proofing or civil engineering groups do various things with the same amount of enthusiasm as the group looking into the water tightness of the earthquake-proof machinery? (November 6)"

Yoshida replied, "They couldn't do anything if the design wasn't determined."

"The machine guys can't design something if they don't know how big a tsunami is coming." He said he then issued instructions to "develop for the interim motors and pumps that'll work even when under water."

However, this idea that "as long as you're not given a design base that anticipates this kind of tsunami, well, then, a true deliberation just isn't possible" or "They couldn't do anything if the design wasn't determined" is far too rigid. Some events incur risks that do exceed the design basis. This is just as certain as the risk of humans making mistakes. Consequently, it is essential to mitigate risk with flexible measures that do not adhere blindly to the design basis.
(Okamoto Akira, Reading the Yoshida Hearings, e-mail to Funabashi, November 28, 2014)

Moreover, behind this logic was a mistrust of people Yoshida referred to as "the safety guys". (He has stated clearly "I don't trust" those "safety guys".) The equipment management unit was divided into a construction section and a repairs section, the "safety guys" referring to people in the construction section. They were, so to say, the safety regulatory authority within the TEPCO nuclear village.

This dislike of the "safety guys" was not confined to Yoshida, but seems to have been shared to a significant degree by engineers from the repairs side, and Yoshida had served as the director of this facility management at Head Office prior to his assignment as site superintendent at Fukushima Daiichi NPS. If he didn't like the way the "safety guys" did things, he should have striven to improve that during his time as director.

While pretending to act in accordance with instructions from Head Office (or top management), TEPCO's nuclear village had actually carried out a different "Kabuki play" several times in the past. The data falsification cover-up that came to light in August 2002 was a typical example. This involved a cover-up by TEPCO's Nuclear Repairs & Inspection Department of 16 incidents of data falsification over cracks in machinery at 13 nuclear plants during the 1980s and 1990s.

There is a standard in the United States referred to as a "maintenance standard", which allows operations to continue in the case of minor damage. Japan, however, requires a stand down in all cases no matter how petty.

A plant's rate of utilization is lowered by an outage. In order to avoid this, TEPCO's maintenance and repairs staff, in cooperation with the manufacturers, checked using their own yardsticks against US standards and even if there were cracks, made false reports of "no abnormality".

The negative legacy of TEPCO's nuclear village looms large in the question of its structural governance.

The Hearings show that no-one starting with Yoshida was free from the yoke of this negative legacy.

Elsewhere, Yoshida claimed that past tsunami predictions were not accurate.

"Who of all the Japanese seismologists and tsunami experts predicted a quake of magnitude 9? The professors researching the Jogan tsunami didn't consider a magnitude of 9 either. I just want to say that if you start saying that, well, you know, that's speaking in hindsight (August 8)."

It was almost as if he was saying there was no need for tsunami countermeasures based on such predictions because the seismologists' predictions were not completely accurate.

This is akin to venting anger in the wrong quarters.

Deeming something is "safe" because the experts say so and the regulators have given their endorsement is nothing but a breeding ground for the "safety myth". Those responsible for the nuclear safety culture have to endure this wisdom in hindsight. They have to prepare for times when in hindsight this occurs.

It is not acceptable if focusing overly on prevention leads to insufficient preparation for emergency measures at the time of a nuclear disaster. The US Federal Emergency Management Agency (FEMA) distinguishes unambiguously between prevention, preparation, and response as the basic concepts of disaster response, and have clarified the necessary measures for each respective area.

In regard to the tsunami measures at Fukushima Daiichi NPS, prevention, preparedness, response, all were inadequate.

Nevertheless, the next remark of Yoshida is beyond comprehension.

"This time, 23,000 people died. Who killed them? They died because a magnitude 9 quake came. If you're going to say that to me, why weren't measures taken then so those people didn't die? (August 8)"

"If you're talking about basic things to protect Japanese lives and property, that's something that has to be taken up in the Central Disaster Management Council and measures put in place by local governments as well. That's where the state is not performing. All it does is talk about nuclear plant design."

What does this mean?

It sounds as if he bears a grudge akin to a sense of victimization along the lines of, since there were no seawalls in the vicinity of the general public's homes, "23,000 people" (actually, the number of dead or missing was 19,610) became victims of the tsunami. Although it was the same for the nuclear plants, why was only the fact that seawalls weren't built around the plants criticized, and moreover, despite the fact that there were no direct victims from the nuclear accident?

Putting damage to the general public and the nuclear plants on the same footing can only be called a truly bizarre logic.

In the first place, residents can be evacuated, but nuclear power plants cannot be moved. After all, Japan is an earthquake-prone country. If you're on the coast, a quake will trigger a tsunami. Yoshida needed to be aware of the dangers of an accident caused by tsunamis, flooding and/or inundation. He was also, in fact, familiar with the flooding accident that occurred in Unit 1 in 1991. This refers to an accident that took place in October of that year when an overflow from a pipe leak caused the emergency D/G to no longer function.

"That stopped almost all of the cooling systems and the D/G was also immersed in water and wouldn't run. I still think that was a terribly big trouble. Barring things that happened this time, that was probably one of the two biggest troubles in Japan although it's not treated like that. That time showed me how scary water can be (November 6)."

If he was going to say that had "showed me how scary water can be", why was the switchboard allowed to remain in the basement of the building?

When facing the afore-mentioned internal review meeting, Yoshida had been told by the person in charge that it would require several tens of billion yen to construct a seawall. Yoshida may have been daunted by the magnitude of the cost. However, tsunami measures were not just about building seawalls. Strengthening the water-tightness of the buildings and the power room, drawing up a manual and training to deal with a full power loss, stockpiling emergency materials and equipment such as batteries and compressors, there were other options.

Although not limited to Yoshida, it has to be said that TEPCO lacked the leadership, organizational culture and the will to share lessons and field experience across the enterprise. It can probably be said that "mental paralysis" led to a rigidity of thinking and action.

The Government Investigation Report (interim report) regarded this point as a problem, voting for the "prosecution" of Yoshida's superiors, Tsunehisa Katsumata (Chairman), Ichiro Takekuro (Fellow) and Sakae Muto (Vice CEO) (Yoshida was excluded due to his death in July 2013).

The committee of the Tokyo District Public Prosecutors Office, which re-investigated in response to the vote, announced on January 22, 2015 for the second time the non-prosecution of ex-Chairman Katsumata and the others (insufficient suspicion).

XII Resident Evacuation & the Role of the Offsite Center

Q: At the time of the Unit 1 vent, in the venting from dawn on March 12, until the vent actually started some time after 9 o'clock, you would have had to coordinate in relation to the evacuation along the lines of there are still some residents yet to leave, do it by the time we evacuate, wouldn't you? This wouldn't have first been done directly in the Emergency Response Center, would it?

A: There wasn't any.

Q: If it was to be done, would it be the office to do it?

A: That's right. By rights, it's the Offsite Center. The basic principles of the Nuclear Disaster Law are that those responsible from the prefecture, the state and the operator go to a place near the site and judge in accordance with events (...) I think the basic philosophy of the Nuclear Disaster Law is that by rights the Offsite Center should decide on what's to be done, but since they wouldn't decide anything, we were told by Head Office and we just made the preparations for the vent, and because when it

> *would be done had evacuation implications (...) Because we heard this and that from Okuma Town, just wait a bit and so on, since things like this happened, rather than shall we wait or we'll wait, it wasn't really a question of waiting or not because we weren't ready, so we just said get out as quickly as possible."*
>
> *"I have no recollection of speaking with the Offsite Center about what should by rights be Nuclear Disaster Law related activities. Funny, isn't it?"*
>
> (August 8)

In the case of a severe nuclear accident, a critically important issue in accident response is how to safely evacuate residents.

In the Fukushima Daiichi nuclear accident, the government decided on the evening of the 11th to evacuate residents within a radius of 3 kilometers, then 10 kilometers early in the morning of the 12th, and 20 kilometers on the night of the in 12th.

The decision itself was taken fairly quickly, but a number of problems remained about notifying and guiding the evacuation. Of these, the greatest problem, which Yoshida noted, was that the Offsite Center wasn't functioning at all.

In the case of a nuclear accident, the Act on Special Measures Concerning Nuclear Emergency Preparedness (Nuclear Disaster Law) required the state to establish an offsite center (emergency response measures base facility) to act as a base for measuring radiation in the locality and collecting information on the nuclear disaster.

Communications systems, nuclear disaster response support systems as well as decontamination rooms in the case of radiation exposure and facilities such as radiation measurement were to be provided.

In the case of a serious accident, the head of the local response headquarters was to use this as an emergency command post. The head of the local response headquarters was stipulated as the Vice Minister of Economy, Trade and Industry. The common offsite center for Fukushima Daiichi and Fukushima Daini was installed in Okuma Town, Fukushima Prefecture. Staff from METI, NISA, the Fukushima prefectural government, neighboring municipalities, TEPCO as well as experts on nuclear power were assembled and sharing information, were supposed to respond to the accident.

On the night of the 11th, the emergency battery at the Offsite Center ran out. That night the only people who had made it to the Offsite Center were six NISA officers, eight TEPCO employees and one staff member from Okuma Town. It was past midnight when Vice Minister Ikeda arrived at the Offsite Center. The Kantei was frustrated that the local headquarters established at the Offsite Center was not working.

At the time, a senior official leading the Crisis Management Center at the Kantei revealed the following inside facts to me.

"A Local Nuclear Emergency Response Headquarters had been set up, but only people with no decision authority, who couldn't say anything to the Ministry proper, were assembled there. Since it was problematic that evaluations had to be made based on dubious information from the prefectural government, we decided to control it from here."

"The Offsite Center had no radioactivity protection facilities, not even air conditioning. The power was shot as well. So, it was useless."

From the night of the 14th until noon on the 15th, everyone from the top (METI Vice Minister) down in the Local Response Headquarters withdrew from the Offsite Center and moved to the Fukushima prefectural government offices.

There was more or less a division od duties between on site, which was responsible for the safety of the reactors, and offsite, which was responsible for the safety of residents and the environment.

However, it became obvious that such a merely formalistic division of roles was not working at all when it came to the Unit 1 vent. Venting involved releasing steam, including radioactive substances from inside the reactor, into the atmosphere in order to lower the reactor's pressure. An artificial division between onsite and offsite was meaningless here.

After the hydrogen explosion in Unit 1, residents within a 20-kilometer radius had to be evacuated in one go. There was no longer any onsite or offsite. After the explosion in Unit 3, all of the staff from the various ministries holed up offsite had to evacuated.

The ultimate challenge in designing nuclear power plant safety is how to design a plan that links together the safety of the reactor and the health and safety of residents at the time of a severe accident.

Although four years have passed since the accident, onsite and offsite response on this crucial point remains unconnected and unlinked, however. While the hard aspects of response have advanced, the soft aspects are sadly delayed.

The Yoshida Hearings show us Yoshida assuming that as site superintendent he could act to evaluate only in terms of reactor logic what to do about the vent. This is based on the belief that the site superintendent has the greatest responsibility for the safe operation of reactors and should devote himself to that.

This stance itself is not wrong, but at the same time, someone with responsibility for both the site and head office needs to respond to offsite issues.

Yoshida himself was also not indifferent to the offsite, expressing his pain over the fact that residents had to live with the insecurity of "how long was this kind of life going to go on?"

However, the assumption that the Offsite Center was going to take care of everything the minute you stepped off the site was highly unrealistic.

The nuclear power business is only tenable to the extent that it can integrate the onsite and offsite and ensure the safety of workers and residents.

Should the impact of an accident reach offsite, operators must carry out aggressive risk management on the basis of the standardized ICS (Incident Command System) code of conduct, involving the field, head office, as well as central and local government.

An approach like this has a span that cannot be matched by so-called forces in field.

The nuclear business should probably be regarded as having "changed to a promise that it will not cause accidents that prevent people from returning to their homes over the long-term."

Based on lessons learned from the Fukushima Daiichi nuclear accident, the government has decided to make evacuation plans for residents within a 30-kilometer radius. The plan is for the government to provide a disaster prevention plan and local governments to formulate evacuation plans, but the reality is that there are no standards or examination procedures to determine the effectiveness of those plans. It would appear that the Government is giving priority to the restart of nuclear power generation without integrating meltdown and radioactivity response with responses that ensure the safety and evacuation of residents.

The Nuclear Regulatory Commission, complacent in its "zen-like" approach of "we haven't said that [nuclear power] is not safe. Nor have we said that it is safe" (Chairman Shunsuke Tanaka), is cautious about involving itself aggressively, saying nuclear prevention measures such as resident evacuation plans are the job of the Office of the Cabinet and local government.

The Regulatory Commission needs to make more of an effort to make resident evacuation plans more realistic and effective.

FINAL REMARKS

The Job of the Government is to Ensure Safety, Not Peace of Mind

"The risk of inaction" and "punishment for inaction"

March 16, 2011 amidst the crisis of the nuclear accident. The Fukushima Labor Bureau sent instructions dated the same and addressed to Site Superintendent Masao Yoshida that, "extraordinary health checks be carried out on workers engaged in emergency work" at the Fukushima Daiichi NPS. Candidates were those whose effective dose exceeded an annual level of 100 millisieverts.

Under the Rules for Prevention of Ionizing Radiation Failure based on the Industrial Health and Safety Act, the maximum level of exposure for nuclear workers was laid down at 100 millisieverts, but in a new directive from the Ministry of Health, Labor and Welfare promulgated on March 15 (and taking force from the 14th) this was raised to 250 millisieverts for "particularly unavoidable emergencies". TEPCO adopted this, tolerating operations by employees who exceeded the 100-millisievert mark.

It was reported in the newspapers at the time that as of 5 a.m. on March 20, the radiation exposure of seven employees exceeded 100 millisieverts (Asahi Shimbun, March 21, 2011).

On the other hand, by the end of March, reports criticizing TEPCO for doing nothing about employees whom they had exposed and the harsh treatment of employees started to appear.

Additionally, Goshi Hosono, assistant to the Prime Minister, also passed on to Yoshida a message from an executive of the US Nuclear Regulatory Commission (NRC) of their "great concern" for the circumstances of the workers at the site.

This was a surprise for Yoshida.

He felt they were exercising due caution with regard to exposure management and had warned Head Office more than once of the risk of worker exposure to large amounts of radiation. In addition, when dealing with the crises in Units 3 and 2, he had implemented a temporary evacuation of staff. In the case of Unit 3, Yoshida evacuated workers for a time to the Anti-Seismic Building on the morning of the 14th when pressure in the reactor was building leading to a possible hydrogen explosion. At the time, Yoshida argued that it was too dangerous, "I can't put people into the field" when Vice CEO Sakae Muto asked him to get workers back into the field for the water injection operations at Unit 2.

When thinking about the treatment of workers, you also need to think about the treatment of the disaster victims forced to evacuate by the nuclear accident.

Yoshida replied to Hosono, "When I think of those people living hand-to-mouth, how can we eat until our bellies are full? You can't just improve treatment of the onsite workers. Improving treatment of the evacuees should come first."

The US side sought several times for an improvement in the working conditions in subsequent US-Japan coordination meetings. At the time, the US side preached, "Three Mile Island showed just how important a nuclear safety culture was in the nuclear power business, the safety of onsite workers also being an important requirement." Even Yoshida would probably not have objected to that point.

At the end of March, preparations for the long haul were put in place with the arrival in Onahama Port of the sailing vessel Kaiwo Maru, a training ship from the National Institute for Sea Training, where onsite workers from Fukushima Daiichi could stay in turns.

Yoshida, however, emphasized during the hearings that the greatest key to weathering the crisis was the field's sense of responsibility and mission during a crisis as well as the strength of its sense of solidarity, and that this was not to be simply taken only from the perspective of a management issue.

> *A: The Tokyo Fire Department's rescue team and so on turned up, didn't it? There was a woman taking care of the firefighters and she was very proactive about her job; she was the woman reported in the newspapers but she suffered internal exposure over the dose. This isn't simply a story of just her being exposed; there were many people who worked with such a real sense of mission and they were exposed; it wasn't just about poor management; I'd like to appeal about how bad the media in this country is.*
>
> *Q: What work was she doing at the time of greatest exposure, this woman?*
>
> *A: It wasn't the work so much that caused her exposure; since the dose was rising in the building itself, by rights the women should have been evacuated early. But she had been working all the time with the disaster prevention group and had a terribly strong sense of responsibility, so we couldn't send her home or rather she didn't go (...) When Unit 1 exploded - the entrance of the Anti-Seismic Building has double doors – (...) at the time of the accident, there were only two doors, so when going in or out, you'd open this one and close that one"*
>
> *Q: She was opening and closing the external doors?*
>
> *A: No, she wasn't opening and closing, but she was standing beside the entrance with the firefighters. In fact, the door was knocked out of alignment at the time of Unit 1 explosion. Even if the door was closed there was still a gap and, so radiation from outside came in; that's how I*

> *suffered internal exposure as well. In her case, she was in the closest place, and so she was exposed internally.*
>
> (July 29)

A technical officer involved in the crisis management at the Kantei at the time, interpreted the true meaning of Yoshida remarks in this passage in the following manner.

"Is this country really right…people making desperate efforts in the field were hurt [by media coverage]. I wonder if a country that can't properly evaluate people engaged desperately in the field can really deal with emergencies… This country doesn't punish people much for doing nothing. There's punishment for things you've done. I think that's a little strange."

In dealing with the Fukushima nuclear accident, the government allowed the permissible annual radiation dose for TEPCO employees to be raised from 100 millisieverts to 250. Nevertheless, even with this standard, the number of people who could no longer undertake operations grew.

There were employees, however, who jumped into the site even when this upper limit was passed. While their actions created a risk, such a "risk of action" is often penalized for breaking the rules. In contrast, no blame is ascribed to the risk of not taking action as long as you comply with the rules.

Yoshida was asking, is that right? How can you respond to a crisis with an approach like that?

Regarding the absence of "punishment for inaction", the flight of NISA officials was just such a case, as was the dysfunction of NISA's ERC (Emergency Response Center) that was meant to act as the secretariat for the Nuclear Response Headquarters.

Another may be added to the absence of "punishment for inaction" list.

The US government provided a contamination map drawn up on the basis of monitoring data to the Ministry of Foreign Affairs, which the Ministry immediately transferred to NISA and the Ministry of Education, Culture, Sports (MEXT), but neither NISA of MEXT made the data public. Nor did they tell the Prime Minister's Office or the Nuclear Safety Commission.

When queried about this later, Deputy Director of MEXT's Science and Technology Policy Bureau, Itaru Watanabe replied, "I think now it would have been better to go public straight away, but at the time there was no thought of taking advantage of the offered data for resident evacuations." He is saying that in the "thoughts" of government officials involved in nuclear power plant administration, monitoring results and resident evacuation was "unanticipated".

The negative power struggle highlighted by the SPEEDI cover-up

When the crisis was at its most hard-fought, the Government did not try to put SPEEDI (System for Prediction of Environmental Emergency Dose Information) to use in the evacuation of residents.

Some of the SPEEDI calculation results may have been too close to the bone. Some of the data may also have been unreliable. There was a possibility that going public would have generated various risks.

The risk that residents in high radiation areas would panic and rush willy-nilly to evacuate; the risk of things getting out of control; the risk of a secondary disaster occurring; the risk of being asked to take responsibility for calculations later found to be mistaken; and also the risk of litigation asking for compensation...

Of all of this, officials were extremely afraid of being made to take responsibility if going public made the public panic.

Regarding this, at a press conference on May 2, 2011 Goshi Hosono, assistant to the prime minister (at the time), replied to the question, "Why wasn't the vent simulation made public?" by saying, "At the time, I think we were afraid of a panic." However, the government officials also in attendance did not admit this and Hosono, under attack by the reporters added by way of explanation, "[the administrators] said something to the effect of wanting to avoid confusion in the general public, that [the fact that panic was the reason] is what I meant. (In the Government Investigation hearings, Hosono regretted his own poor crisis communication, "Because it was perceived that they had hidden [the SPEEDI calculations] out of panic. That was a real failure on my part.")

They elected to take the "risk of not acting" over the "risk of acting".
It was the Ministry of Education that clung completely to the "risk of not acting".

The SPEEDI cover-up by the Ministry of Education during the Fukushima nuclear accident was nothing more than the negative power struggle characteristic of the Kasumigaseki bureaucracy.

A negative power struggle refers to the Kasumigaseki art of never raising your hand, never standing forward, and always remaining inconspicuous for anything that will not win you political points, that was not a plus in terms of official power, that would reduce parachute posts, that forced you to take on onerous jobs, or that interfered with the career advancement of top officials.

In the midst of the crisis, the Ministry of Education orchestrated an attempt to unilaterally transfer the jurisdiction of SPEEDI to the Nuclear Safety Commission. Perceiving that SPEEDI would turn into a political issue, it tried to force this onerous job onto the Nuclear Safety Commission.

There is a view among the faction critical of SPEEDI that "it was a failure as expected that in the first place SPEEDI wouldn't be of any help in a pinch."

Claiming that SPEEDI is supported "because it suits the various parties", Seiji Abe (Technical Advisor to the Technical Management Officer at the Secretariat of the Nuclear Regulatory Agency) gave the following interpretation.

- Many of the national and local government disaster prevention personnel did not have sufficient knowledge of what comprised a severe accident. Of course, in dealing with the accident they were unable to determine about withdrawal and/or evacuation. All they had to say was "It's OK, we have SPEEDI."
- Even if you know that it is not useful in disaster prevention, as long as you do not profess that, huge development and improvement costs as well as SPEEDI related equipment maintenance costs flow in.
- Because it is bothersome for the media to come to grips with technical issues, they can easily carry on with a "they hid it, they hid it."
(Abe Seiji, Nuclear Risk and Safety Regulations: Before and After the Fukushima Daiichi Accident, Daiichi Hoki, 2015)

To conclude, it is questionable as to whether the SPEEDI simulations could be used as expected as specific grounds for resident evacuations in the midst of the Fukushima nuclear accident.

However, we need to remember the "inaction" of the government, especially the Ministry of Education which has poured 13 billion yen into developing this system, of only delivering words and deeds that made light of its use and utility at a time when it should have been center stage, as well as not releasing the data.

The phrase "in order to avoid panic from residents" is often none other than another name for risk aversion.

Unless "the risk of inaction" and "punishment for inaction" are consciously incorporated into risk management and crisis management, in the case of a future crisis, organizations are likely to exacerbate the crisis by taking the "risk of inaction" (in other words, risk avoidance). And not imposing a "punishment for inaction" will only encourage no one to take responsibility or learn the lessons.

"Overly seeking a small peace of mind leads to neglect of the greater issue of safety"

The reason politicians and government officials fear to this extent "avoiding panic" by residents was because they had been pursuing a hard sell of not only promising "safety" but also "peace of mind" to residents. This derived from the paternalistic governance of the government, which just as parents, treated citizens like children.

Shichihei Yamamoto, a theorist on issues of Japanese national and cultural identity who had once taken the world by storm, has written in his book Why Japan Was defeated: 21 Causes of Defeat that the Japanese "try to replace realistic solutions with psychological solutions."

Having been through hell at the Philippines Front as an artillery lieutenant, Yamamoto's theory of Japanese society includes insights into the issues of risk, governance and leadership in the Japanese social system that could also be dubbed "nuclear defeat", his observation that we "try to replace realistic solutions with psychological solutions" being a penetrating critique on the psychological, social and political structural issues that try to force via "inner guidance" the acceptance of a government endorsed "safety" by expanding for no reason all regulations and the establishment of a safety culture to encompass "peace of mind" instead of setting its sights on "safety" alone.

It is easy for risk assessment to become ambiguous when safety is wrapped in a film of peace of mind, and when you attempt to turn the realistic solution of safety into the psychological solution of peace of mind. It is difficult in this kind of context to bring risk to the forefront, and to coolly evaluate the probabilities and a trade-off between the "risk of taking action" and the "risk of not taking action" and to decide on an acceptable range of risk.

It is easy to be tempted to try and change the direction of risk assessment when you are trying to carry through risk management in terms of cost, internal procedures, management, resident measures, the media, and politics. This can lead to a tendency to fall into a negative stance of making the "risk of taking action" more problematic that the "risk of not taking action."

It is in this way that the tendency for "overly seeking a small peace of mind leads to the neglect of the greater issue of safety" grows stronger in both politics and government administration. (It was Nobukazu Niigata's Why Don't the Japanese Try to Think?, Shinyosha, 2014, that familiarized me with Shichihei Yamamoto's ideas.)

When you look at the Japanese situation of being unable to start out towards rebuilding from its "nuclear defeat", questions such as is there something unique in the Japanese system, does the problem lie in Japanese culture itself, spring to mind.

A typical view of this approach is probably the theory of Kiyoshi Kurokawa, Chair of the Diet Investigation Commission, that the Fukushima nuclear accident was "Made in Japan".

In the abridged English version of the Diet Investigation Report on the Fukushima nuclear accident, Chairman Kurokawa sums up the disaster as Made in Japan, pointing out that it lay in "our reflexive obedience, our reluctance to question authority, our passion for 'seeing a plan through', our collectivism, and our isolative island mentality."

However, as is the case with historical determinism and ideology, interpreting the essence of the problem though Japanese cultural theory (or a theory of the Japanese) and focusing on the characteristics and qualities of the Japanese people and Japanese society as a whole is highly problematic as an intellectual framework.

To begin with, these cultural theories or theories of Japanese identity are conveniently trotted out as an explanation whenever Japan is in a steep descent or ascent. Even so, it is possible for the historical and structural background to Japan's bureaucratic and corporate culture (institutional culture) to define the behavior and philosophy of individuals in the organization and constrain the risks, governance and leadership required during crisis response, and as a result, give structure to an "essence of failure".

At such a time, it cannot be denied that the cultural characteristics and psychological tendencies of Japanese society embalm Japanese organizational culture like an amniotic fluid and seep in.

It is quite likely that the "one-country safety" approach and inertia that formed the backdrop to the delay in making severe accident response compulsory, in international cooperation on countermeasures for nuclear terrorism and international coordination over the marine disposal of contaminated water are born of this underlying organizational culture of Japan.

To rephrase it, the historical and structural factors behind the nuclear accident were not "Made in Japan" but "Made inside Japan" (made to Japan specifications).

Another point is that relying on Japanese culture determinism makes it difficult to "learn from failure".

- If it is culture that is at fault, the role and responsibilities of the individual become abstracted. When it becomes the responsibility of the whole it is no longer anyone's responsibility.
- It leads to defeatism along the lines of any effort is useless because you can't change culture.
- If it becomes a question of Japanese culture, the event becomes unrelated to the world at large. Investigating the causes and sharing lessons with the world is meaningless.

Who from where, what part of which organization, under what circumstances and in what structure, what action taken by what judgment under what calculations brought about what kind of result?

It is imperative that each of these questions is researched and verified scientifically one by one. The answer should not be dumped wholesale into the lap of national culture. (Please refer to my book Nuclear Defeat - Crisis Leadership, Bunshun Shinsho 2013, on this point.)

To conclude, Japan's state, its society and organizations, failed to properly take into account the issues of risk, governance, and leadership; to face them front-on; that is the essence of the problem.

The Fukushima nuclear accident specifically highlighted the following the issues.

- Refusing to entertain a "worst-case scenario", safety was replaced with peace of mind and everyone bunkered down in the safety myth. Preparations, logistics, training were turned into "mere form" (making risk taboo).
- Insiders bumbled their way through, different points of view, stances and groups all being excluded. Day in, day out it was silos, foxholes and turf wars. They were no good at polishing their ability to oversee the whole system and to create a system where everyone "pulled together" (lack of horizontal governance).
- Failing to clarify authority and responsibility, they couldn't establish a chain of command (lack of vertical governance).
- They were unable to set clear priorities including cutting losses (lack of leadership).
- They could not create "Big Politics" where the "nation united" in a time of crisis (lack of leadership).
- They were no good at paving the way to resilience (recovery) by proactively disclosing the risks, acknowledging failure as failure, then getting back on their feet (lack of leadership).

At the same time, there is a need to know how difficult it is to learn from previous examples. Yoshida himself touched on this point taking Kashiwazaki NPS during Chuetsu Offshore Earthquake of 2007 as his case.

"The Chuetsu Offshore Earthquake in Kashiwazaki (...) In short, it shutdown operations safely (...) I mean, even with an earthquake that size, it stopped properly, didn't it, and when they checked later, it was way past the level of earthquake it had been designed for, and yet almost all of the safety equipment was intact. (...) Since that was proved in a certain sense even in an earthquake that far exceeded the designed ground movement, I did feel conversely that the Japanese design was right. (November 6)"

Events that surpass the design do occur.
However, it escaped unscathed because they had responded flexibly and taken a margin over and beyond the design.

They were laid low by an "unanticipated" event, however, because they saw the success at this time as meaning the design was correct and put the form of the design on a pedestal.

That was probably what Yoshida was trying to say.

(Some honorific titles have been omitted from the text)

[PART 2]

Three Years on from Publishing The Independent Investigation Report

(An analysis and investigation three years after the event by four members of the then Working Committee)

Kenta HORIO
Akihisa SHIOZAKI
Kazuto SUZUKI
Shinetsu SUGAWARA

Has Japanese Society Learnt the Lessons from the Accident?

Kenta HORIO

Almost four years have passed since the crisis of the Great East Japan Earthquake and the accident at the Fukushima Daiichi Nuclear Power Station, unprecedented in the post-war era. I participated in the Independent Investigation for about half a year from the summer of 2011 with a mind to understanding why the accident had happened and what was the nature of the nuclear accident we had faced. Today, some three years since our report was published, I would like once again to look back at the nuclear accident and our investigation.

Researching/Analyzing the Accident and the Role of the Independent Investigation

During this time, there has been significant progress in the investigation and analysis of the accident. At the time we issued our report, the only main sources of information about the accident that had been published were the Japanese government report to the International Atomic Energy Agency (IAEA) immediately after the accident, the Interim Report by the Government Investigation Committee, and the interim report of the accident investigation committee that was established in-house at TEPCO, the party directly involved in the accident.

However, in February 2012 after publishing our report, this was followed by the compilation of the Final Report of the Diet Investigation Commission as well as that of the Government Investigation Committee, in addition to an accident investigation report compiled by the Atomic Energy Society of Japan. Apart from its final report on the accident investigation, TEPCO also published a "Comprehensive Nuclear Safety Reform Plan

Part 2   Three Years on from Publishing The Independent Investigation Report

for the Fukushima Nuclear Accident", which looked more closely at the cause of the accident.

Furthermore, TEPCO and the Nuclear Regulatory Commission, which was established after the accident, each conducted a further investigation and analysis with two previous documents, a "Progress Report on the Investigation and Examination Results on Unconfirmed/Unresolved Matters (TEPCO) and an "Interim Report on the Analysis of the TEPCO Fukushima Daiichi Nuclear Disaster" (NRC), made public. The hearings records of more than 200 people, including the Government Investigation hearings with the (then) Site Superintendent Masao Yoshida (the Yoshida Hearings), have also been released since last fall.

The public release of several accident Investigation Commissions and their follow-ups as well as the hearings records that form the cornerstone for verification have made it possible to mutually compare a variety of research and investigation results and through this re-verification, to draw more robust lessons. The role the Independent Investigation played can also be placed in this context, and in particular the early stage at which its report was published and its efforts to describe the facts in as unembellished manner as possible provide a "yardstick", I believe, for deciphering the Government and Diet Investigations.

The Independent Investigation as seen from the Yoshida Hearings

The strength of the Independent Investigation was that it conducted its investigation autonomously from TEPCO and the Government, both involved in the accident, as well as politics, using an approach that barred no taboos regarding a variety of matters relating to the accident. On the other hand, one of its weaknesses was that, since it had no legally binding authority, it was unable to survey TEPCO parties, especially those struggling onsite to respond to the accident from Site Superintendent Yoshida down.

Facts concerning how the event developed and the responses taken onsite were, of course, confirmed, analyzed and discussed based on all the available information at the time such as the interim report of the Government Investigation. I don't believe there were any major discrepancies in those results. However, regarding the depth of analysis or perspectives on the lessons to be learnt, it is also true that there were limitations. Citing some iconic remarks from the Yoshida Hearings below, I would like to point out two lessons that seem particularly important but are missing from our report.

Support from Offsite:

> *"You know I'm still tremendously bitter about the fact that we didn't receive any substantial or effective rescue"*
> (July 29, 2011)

The Yoshida Hearings demonstrate repeatedly dissatisfaction with support from outside the site, this confessional remark from the field commander, albeit prefaced by "I'll speak of my feelings", being a weighty one.

It is not as if, of course, both Head Office and the Kantei failed to provide any support during the accident. For example, Head Office looked into such things as assessing the balance between the temperature of decay heat and the amount of water injection as well as the environmental impact of a vent; technical assessments needed for onsite decision-making; and methods for continuous water pumping into the spent fuel pools (August 8). However, in most cases, Yoshida's assessment was severe: "they made so many queries" "it wasn't support" (July 29).

Logistics also came in for the same harsh assessment. One of the features of the accident response was the fact that onsite resources alone could not support the response, but equipment and supplies from outside were required. However, during support from Head Office and elsewhere, no consideration was given to transportation means or sorting and the fact that it was necessary to devote onsite manpower to these jobs was described as a "huge loss".

The fact that Head Office and the Kantei could not necessarily provide effective support for the onsite response, and that field command was not satisfied with the response of Head Office and the Kantei had also been recognized when writing our report. However, while the Independent Investigation mentioned the government scientific advice function, it did not cover support for field command due to a lack of primary information.

Of course, it is not my intention to simply defend Site Superintendent Yoshida, but reducing the stress of field command and preparing response measures as much as possible should be borne in mind for accident response in harsh environments such as this.

There is also a fine line, however, between "effective support" and "excessive intervention". Taking into account the possibility of differences in awareness occurring as is described later, there is a strong danger of falling into excessive intervention regarding support especially for decision-making. Therefore, while it is difficult to provide an appropriate yardstick for what needs to be done to be able to say that lessons have been sufficiently learnt, at the very least we should have placed more emphasis on the factor of external support, especially "support for field command".

Differences in awareness, communication:

> *"Differences in awareness between the real field of the Central Control Room and the quasi-field of the Emergency Response Center and the far-removed Head Office became glaring."*
>
> (July 29, 2011)

Another point I wish to highlight is the issue of differences in awareness and communication, which the above quote from the Yoshida Hearings expresses succinctly.

Although the Independent Investigation report also pointed out the fact that a "multi-layered information pathway" acted to complicate the accident response, our comment was focussed more on the information pathway from the site to the Kantei. However, Site Superintendent Yoshida frequently mentioned there was a difference of awareness within the site as well and additionally between the Central Control Room and the Emergency Response Center.

One of the points Yoshida himself cites as a point of regret especially is the response related to the IC at Unit 1, looking back that "the SOS the field side sent didn't reach us" (July 22). Given that the IC was a key piece of equipment in responding to a severe accident, it left a lesson in terms of design and instrumentation, but it also offered lessons in terms of sharing information and awareness in harsh environments.

The fact that physical means of communication were extremely limited was a major cause of the difficulty that occurred in conveying information and communication during this accident, but it should be noted that "information" in this case refers not only to comparatively objective facts such as progress reports on jobs at hand or instrument readings, but also to uncertain or subjective information such as predictions about future situational developments, or concerns and a sense of crisis.

Therefore, even if the means of communication and so on had been sufficient, it is unlikely that the differences in awareness that Yoshida pointed out would not occur at all, the lessons to be learnt probably including the fact that accident response training should take into consideration not only ensuring the means of conveying information, but also the fact that difference in awareness occur. This is also important in terms of the afore-mentioned "field command support".

Where have the lessons been reflected?

As was pointed out at the start, certain progress has been made over the past four years in investigating and verifying the accident, and as a result various lessons have been deduced. So, has Japanese society learnt the lessons from the accident? While somewhat fragmentary, I would like to try to answer this question.

The general answer is yes and no.

For example, the many investigations including the Independent Investigation have pointed out shortcomings in deep defense at the fourth layer (severe accident measures) and the fifth layer (disaster prevention), especially the (increasing isolation) discrepancy with international standards. The reform of nuclear safety regulations after the accident has seen regulatory requirements established for severe accident measures and a start to the introduction of probabilistic safety assessment methods, and even in disaster prevention, decision-making and preparation areas for resident evacuation have been revised in line with international standards. Additionally, regarding shortcomings in the independence and expertise of the regulatory agencies, which was cited as a background factor, the Nuclear Regulatory Commission (and the Nuclear Regulatory Agency) established after the accident is much more independent than before, the system now capable of performing more centralized administrative regulation.

Elsewhere, safety measures including severe accidents still need to be constantly reviewed in the light of new knowledge, and disaster prevention requires ongoing efforts such as the establishment and improvement of actual operation performance. Moreover, various opinions abound even now concerning the state of the regulatory authorities, the lesson there being that sustained efforts are required for training highly specialized regulatory staff.

In terms of the system for nuclear power safety legislation and organizations, it can therefore be said that based on the lessons learned from the accident the situation is approaching what it should be, but there are still areas including raising effective capabilities that require continuing efforts in the future. Additionally, consistency with international standards will require improvements based on the results of the comprehensive nuclear safety regulation evaluation service (IRRS) by the International Atomic Energy Agency (IAEA) that Japan has accepted to undergo.

Of the lessons to be learned, there are also some areas where who is to mainly do the learning is not necessarily obvious. For example, this concerns the lesson of "ambiguous national policy/private management", in other words clarifying the public and private sectors' role (responsibility) sharing. Given that based on power system reforms, changes in the relationship between public and private sectors in the nuclear power business can be expected, attention needs to be paid to how that role sharing is reviewed and what mechanics come into play. In addition, the "myth of absolute safety" that makes risk visualization a taboo needs to be overcome at various social levels, not only in operators and regulatory agencies, but in central and local "nuclear villages" as well as outside those "nuclear villages". However, it is no easy matter to ascertain properly if lessons are being learnt not as a question of a single individual's ethical views or a single organization's organizational culture. Moreover, even after breaking away from the "safety myth", higher order challenges such as the extent of uncertainty to be considered or what sort of risk should be responded to in what time frame will remain.

Given these points, it would not be appropriate to mischievously deem that the lessons of the accident have not been learnt. Nevertheless, it is also a fact that ongoing lessons and challenges remain to be learned in the future, a point itself that must be continually taken to heart.

Part 2　Three Years on from Publishing The Independent Investigation Report

The need for and importance of further verification

As a final point, I would like to touch briefly on what needs further verification in the future. Based on the results of the research, analysis and verification on the accident I conducted in the past, I believe they should be acknowledged to a certain degree, but at the same time it should be recognized that there are areas that requires further verification.

The first is something that TEPCO and the Nuclear Regulatory Commission have already carried out and that is the additional investigation of matters where the opinion of the various investigations diverge or remain unclear. In particular, concerning the situation at the site, as the decommissioning work makes progress and a greater area can be accessed, new findings are likely to emerge and a constant updating of the accident analysis and lessons learned based on new knowledge will be required.

A second factor is areas that did not enter into the scope of the various investigations, but require a longer span of investigation. The best examples of this are dealing with the nuclear power plant that caused the accident (decommissioning), and decontamination (environmental restoration) as well as the recovery and reconstruction of the surrounding area. These challenges cannot necessarily be verified now, and although difficulties can be expected concerning who will carry out this verification and using what approach, its importance as the subject of verification is beyond discussion. As such, I believe it is important to keep the possibility of verification as open as possible for future generations.

Furthermore, I also believe that expanding somewhat the historical perspective that the Independent Investigation emphasized for an even deeper consideration of nuclear energy policy and nuclear administration or the historical development of the nuclear power business is also required.

The author experienced the accident when he was at graduate

school, and after participating in the Independent Investigation, has been working from the spring of 2013 as an expert investigator at the Japanese Mission of an international organization in Vienna. Although not currently involved directly in issues relating to nuclear safety (especially domestic safety regulations and safety measures), to investigate the Fukushima nuclear accident and to engage in the issues that emerged from the accident is the responsibility of we who are involved with nuclear power, and I would like to maintain this stance no matter what my position. I am not sure if I have provided a sufficiently deep discussion as a "re-verification" to readers, but will be pleased if I have played a small part in continuing to learn from the accident.

Who Should Put Their Life on the Line to Defend a Nuclear Power Plant?

Akihisa SHIOZAKI

> A: If you'd been the site superintendent in the same place at the same timing, I think you'd probably understand, but when you have three units going wild like this and all sorts of information pouring in, when you have to make a judgement call, you just don't know what's going on anymore. So, my instructions were, what I instructed was simply get the water in, do something and get the water in, I don't care if it's seawater, and get the pressure in the containment vessel down
>
> (July 29, 2011)

The crisis response in the Fukushima nuclear accident cannot be discussed without referring to the leadership of Site Superintendent Yoshida at the disaster site. As General Manager of the Emergency Response Headquarters at the Fukushima Daiichi Nuclear Power Station, Yoshida spearheaded the disaster response in the field, eventually stabilizing a serious nuclear disaster such as never before experienced. The Yoshida Hearings reveal Yoshida's strong sense of mission and responsibility in the face of one difficult decision after another during extreme confusion and time constraints, together with a vivid account of the loneliness and frustration of his lone struggle in an environment isolated from the outside and contains many lessons.

The long-term loss of all power due to the tsunami following the large-scale earthquake was an event that far surpassed any anticipated nuclear disaster prepared for by TEPCO in normal times. Moreover, the operating procedures for the accident assumed that the state of the reactor could be monitored through parameters, but if, in the first place, the parameters could not be read due to a total power loss, those procedures became worthless as a guide for the field command's crisis response. Yoshida and the others had no manual to depend on, and under conditions where situational awareness through instrument measurement was not even possible, they had to confront the unimaginable task of simultaneously cooling five reactors, where a rapid rise in temperature was taking place, unarmed so to speak.

The Yoshida Hearings depict the nature of the struggle by Yoshida and his colleagues, searching by trial and error for some way to improve the situation in an "unanticipated" scenario. For example, at the time of the Unit 1 vent in the early morning of March 12, Yoshida tried to get his subordinates to open the valve manually because there was no power supply but they failed because the dose was too high. They tried to raise the pressure by bringing in a compressor laden with batteries and open the valves again by remote control. Also, when it was decided to inject the seawater into Unit 1, since there was no booster to raise the water from the sea surface to a height of 10 meters, they demonstrated their ingenuity by diverting the seawater from the tsunami accumulating in the backwash valve pits and turbine of Unit 3, followed by lining up fire engines in two rows to take the place of booster and lifting the sea water.

In this way, in the midst of their struggle with an ever-changing disaster situation, the field implemented one creative and ingenious idea after another from Site Superintendent Yoshida . In a desperate situation and without fear of failure, the positive attitude of the site of doing whatever was possible is worthy of certain praise as a crisis response, irrespective of individual success or failure.

Part 2   Three Years on from Publishing The Independent Investigation Report

Distrust of poor support and interference

On the other hand, it is not the case that Site Superintendent Yoshida reported or consulted closely with Head Office about all of these onsite impromptu decisions. For example, regarding investigations into ways of injecting seawater, Yoshida replied that he "cut the [video] voice link"[1] since there was no prospect of particularly useful ideas being put forward even if he did ask Head Office, and he did not necessarily seek Head Office's opinion when he was considering individual responses, but professed rather that the site led the advance. And Yoshida statements all suggest that it was the high level of capability in the field and their positive attitude as well as a distrust and dissatisfaction with the poor effective specialist support from Head Office that was behind this site-led approach.

In the first place, Site Superintendent Yoshida revealed his disgust about the fact that the operational standards for Fukushima Daiichi NPS did not assume a total power loss and that they provided virtually no guidance for onsite response, saying, "Those idiots, they were of no use whatsoever."[2] Additionally, Yoshida asserted that the advice and instructions from Head Office merely fanned anxiety and impatience and were of no use in responding to the problems as illustrated by his rebuff on the 14th, "Please stop asking me all this stuff. We're trying to get the containment vessel vent open now. Don't disturb us."[3] Yoshida did not hide his irritation at the lack of substantial support from Head Office saying, "There were many queries. What's the situation now and so on. That wasn't support. They were just asking so they could report back. I lost my temper halfway through and remember telling them to pipe down or shut up more than once."[4] And as symbolized by Yoshida's remark, "I'm still tremendously bitter about the fact that we didn't receive any substantial or effective rescue whether it be from Head Office or wherever despite all those people doing all that stuff"[5], the scene suggests a state where the site was left behind in a deep sense of disconnection and isolation from the outside.

Moreover, as mentioned in the Independent Investigation

Report, in the initial response to the Fukushima nuclear accident, there were several instances of direct interference with the site from the Kantei, which led to confusion. At the same time as strongly criticizing this interference from the Kantei, Site Superintendent Yoshida said, "Why's the Kantei coming here directly? What's the headquarters at Head Office doing?!",[6] showing his strong dissatisfaction with Head Office for not acting as an effective buffer against interference from the Kantei. The description in the Yoshida Hearings suggests that inside TEPCO was by no means monolithic, but a complex and tense relationship existed between the field side struggling while being exposed to life-threatening danger and Head Office struggling to provide support sandwiched between the site and the Kantei.

Collapse in the chain of command

Yoshida's sense of distrust in and isolation from Head Office and the Kantei created a serious problem in the chain of command for the Fukushima nuclear accident response. The most symbolic scene was the series of circumstances surrounding seawater injection into Unit 1 on the evening of March 12.

At the site when the seawater injection at Unit 1 had started at 19:04 on the same day, there was a phone call from TEPCO Fellow Takekuro at the Kantei to Site Superintendent Yoshida at 19:25, asking them to stop the seawater injection at Unit 1 until the prime minister's approval had been obtained. After consulting with Takahashi at Head Office, Yoshida decided to report that he would follow the Kantei's instructions. However, regarding the fact that TEPCO had already started a partial seawater injection at this time, the situation was handled differently than the actual state of affairs, it being decided to say that this was only a test injection to check the pump lines. Regarding the expression "test injection", Yoshida stated "I think it was Head Office that came up with that,"[7] claiming it was the Head Office side and not himself who devised this expediency. That is, fearing it would create a problem if they reported to the Kantei that a situation different to the wishes of the Kantei was already being pursued at the site, Head Office neglected to provide an exact status report to the Kantei with Yoshida's cooperation.

On the other hand, Site Superintendent Yoshida ignored instructions to stop from the Kantei and Head Office and continued pumping in seawater. While declaring a stop to seawater injection at the Roundtable, he instructed the disaster prevention group leader, "I'll give the order to stop, but you're absolutely not to abort," continuing to pump water secretly in violation of his instructions. Yoshida commented on his feelings at the time, "But since there was no way I intended to get rid of the water at this time, to stop the injection (…) since I couldn't follow any instruction that offered no collateral on when we could resume, I decided to act on my own judgement."[8]. His tone suggests that against the background of the above-mentioned distrust and dissatisfaction with the Kantei and Head Office, Yoshida believed it was field command that had the final authority regarding disaster response instructions from the Kantei and Head Office.

Complying with the chain of command is an absolute in crisis response. As the Independent Investigation Report pointed out, it has to be said that one false step and the inaccurate status reports from Head Office to the Kantei and Yoshida's infringement of the directive to stop were actions fraught with danger that could have escalated the disaster further. This time these actions did not result in major damage, but the evaluation of this violation of instructions, the excessive interference from the Kantei that led to it and the tension that existed between the field and Head Office, these all need to be thoroughly examined to prevent any recurrence.

Relying on individual qualities in a nuclear disaster response

It is impossible to perfectly predict or anticipate all future nuclear disasters. No matter how much safety measures are strengthened, no matter how detailed a crisis response manual is drawn up, unanticipated developments will always occur, and it is highly likely that much of the response will rely on impromptu decisions in the field. The possibility of a recurrence of tension between head office and the government may flare up and split views on serious decisions cannot be denied. Each statement in the Yoshida Hearings vividly relates how the outcome of the country's nuclear disaster response relies on the individual qualities of the field command.

Improving and standardizing the crisis response capabilities of each nuclear power station site superintendent is one of the most crucial issues for nuclear disasters in Japan. Nuclear site superintendents must be selected with an emphasis on their suitability for crisis management, including greater than ever stress tolerance and judgment in advance anticipation of having to face a serious nuclear disaster. Unfortunately, Site Superintendent Yoshida, who had the best understanding of the lessons from the Fukushima nuclear accident, is no longer with us. Luckily, however, many of the hardships experienced by Yoshida and the resulting lessons learned have been left on record. These should be taken advantage of to ensure that the head office side can provide the site with more valuable advice along with the implementation of thorough training and education during normal times on the technical aspects of crisis response, including the IC and IRRC mechanisms. This is not an issue that should be left up to the individual response of regional power companies. It is crucial that under the cooperation of both the public and private sectors a nation-wide standardization of the crisis management capabilities in nuclear power plants be achieved.

> *A: At the time, I called in so-and-so from personnel affairs, and quietly into a room at that, and told him to check how many people there were. (...) I told him to check particularly the number of people not involved in operations and repairs. Not that we could do anything about the people assigned from Head Office. How many buses could we use? I thought there were two or three, but what about*

Part 2   Three Years on from Publishing The Independent Investigation Report

> drivers? Were they fuelled up? I told them to have them wait out front. Although it hasn't been mentioned here, I made preparations so that they could move out and evacuate immediately if something happened. Those were my instructions
>
> (August 9, 2011)

TEPCO's evacuation request and the Kantei's response

The response of the Kantei and TEPCO from the night of March the 14th through to the early hours of the 15th during the establishment of the integrated headquarters is one of the scenes where evaluation was most divided in the series of investigations on the Fukushima nuclear accident.

Specifically, after the evening of the 14th, owing to confirmation that the fuel in Unit 2 had become exposed and the radiation dose was on the increase, the TEPCO side began looking into a withdrawal from Fukushima Daiichi. When, from the evening of the same day until late at night, TEPCO CEO Shimizu placed one phone call after another to consult with METI Minister Kaieda, Chief Cabinet Secretary Edano, Prime Ministerial Aide Hosono and NISA Chair Terasaka, the Kantei interpreted this as a request from TEPCO for a "complete withdrawal". Learning of these developments, Prime Minister Kan called CEO Shimizu to the Kantei around 4 a.m. on the 15th, refusing a "complete withdrawal" and subsequently stormed over to TEPCO, leading to the establishment of an integrated response headquarters, including the permanent assignment of Hosono to TEPCO Head Office.

The Independent Investigation Report evaluated this response positively, writing "Prime Minister Kan's refusal of a TEPCO withdrawal was not necessarily accompanied by specific measures aimed at stabilizing Unit 2, but was mainly based on a strong sense of crisis that a withdrawal would surely aggravate the situation. However, it was a turning point in the crisis response in that the refusal forced TEPCO into a stronger resolve, and acted as a catalyst for establishing the integrated response headquarters at TEPCO Head Office."

The Independent Investigation did not always obtain sufficient research cooperation from the TEPCO side and reserved judgment on whether the TEPCO side's real intent was a "complete withdrawal" or a "temporary/partial evacuation". Later, however, as progress was made in the Government and Diet Investigations, the reality of the TEPCO side's request on the night of the 14th became increasingly apparent.

According to Kunio Yanagida, who served as a member of the Government Investigation, in the verbal exchange during a site visit on June 30, 2011, Site Superintendent Yoshida vividly described the extreme situation they had reached from the night of the 14th to the morning of the 15th. More specifically, he said Yoshida stated, "we were finally able to lower the reactor pressure (in Unit 2), the hoses had also been reconnected, and we were in a situation where water could be injected. Just as we breathed a sigh of relief and were about to start pumping in from the fire engines, the fire pumps wouldn't budge. They'd run out of fuel. And what's more, they said they couldn't get hold of fuel straight away. That was the only time I thought we'd had it. Things were in a panic. If Unit 2 exploded, because people wouldn't be able to stay on site at Daiichi, not only would we not be able to control Units 1 and 3 on either side, but they would have to stop all work at Daini due to radiation, and so it would be a real mess." (Bungei Shunju, November 2014 issue). And the feelings of Site Superintendent Yoshida at the time in the midst of these circumstances were, "This's no good. Something like, now's the best time to die (ibid.)". He had been cornered to the point of preparing to die.

Under these circumstances, from the evening of the 14th, preparations were afoot at Tokyo Head Office and the site for a large-scale "evacuation" plan to evacuate the majority and leave behind only those essential to the control of the reactors. By the night of the 14th, there were still over 700 people at Fukushima Daiichi NPS including personnel from associate companies.

Although it is not necessarily clear from objective documentation how many people the TEPCO side planned ultimately to leave on site if the situation deteriorated, Site Superintendent Yoshida Director states in the Diet Investigation Report, "It was difficult to determine how many people should be left, we weren't thinking in terms of numbers at that point. Nevertheless, I thought there might be ten or so people I knew from way back that would die with me at the very end," acknowledging that at least for him, it was his intention in the case of a sharp jump in the radiation dose to get the majority of staff with the exception of a small "death squad" out of Daiichi although it wouldn't be a full withdrawal.

The lack of a "worst scenario" response plan

Regarding this question, the focus of media coverage up until now has been whether the TEPCO side had truly intended a "complete withdrawal" and if that was not so, who at the Kantei was responsible for this misunderstanding. A more serious issue, however, in the context of an ultimate national crisis where many parties prepared for the "devastation of Eastern Japan" triggered by the Unit 2 meltdown, is verifying the cause of such a serious miscommunication between the Kantei and TEPCO and preventing a relapse.

What was revealed accidentally in the course of the afore-mentioned confusion from the night of the 14th to the early hours of the 15th was the absence of a plan on either the side of TEPCO or the government to deal with how to respond to a crisis that threatened both life and body in the "worst-case scenario" of a nuclear disaster where the radiation dose was climbing high enough to cause serious adverse effects on the health of site workers. In other words, there is no evidence of a priori discussions between the government and TEPCO as to whether in the case of a sharp jump in radiation, crisis response capability was to be maximized even at the risk of endangering the health and lives of site personnel, or priority was to be given to protecting the health and lives of onsite workers even if this meant sacrificing crisis response capability somewhat, and no manual existed that defined field evacuation philosophy or procedures in such a situation. This is ostensibly one key reason why it was only under the instruction of CEO Shimizu on the evening on the 14th that a stopgap evacuation plan was developed on the TEPCO side, and although staff inside TEPCO tried behind the scenes to obtain approval from the Kantei in advance as the work of drawing up a plan was going on, the afore-mentioned misunderstanding, distrust and miscommunication came into being.

Even now some four years since the Fukushima nuclear accident, national debate on this most important and difficult issue of who should risk their lives in responding to a nuclear disaster when radiation has increased to a level that has a seriously adverse impact on health and life is not any deeper.

To put it extremely, Prime Minister Kan's refusal of a withdrawal and his subsequent behavior in the early hours of the 15th can be interpreted as one approach of "power company personnel should do their utmost to respond to a nuclear accident to the very last even at the risk of their own lives, and are not allowed to leave the site." In other words, this can be taken as meaning whether it be a "complete withdrawal" or a large-scale "temporary evacuation", in a national emergency such as a severe accident at a nuclear power plant, the maximization of crisis response capability in a disaster should be given the highest priority, and even if it was a private business such as TEPCO, even if it endangered the lives and bodies of personnel, they should offer their full cooperation for the above purpose. Of course, the dissenting opinion also exists that it is not reasonable to inflict such a heavy responsibility on a single private company like TEPCO and its staff; that in such a case the government should address it. This would mean public officials such as the Self-Defense Forces and fire fighters responding as the execution unit to the crisis, therefore requiring the implementation of courses and training on expertise in nuclear disaster response. Furthermore, at a stage where the dose has risen to a considerable degree, an approach where onsite response, whether it be public or private, is abandoned and all-out evacuation pursued to preserve human life as the top priority is also conceivable. In such a case, the government would be required to disseminate widely and explain to the public in advance the serious risks inherently entailed in the infrastructure of nuclear power plants.

Additionally, attention needs to be paid to the procedural aspects of who in an actual serious crisis scene will make such difficult decisions. Based on contrition over the Fukushima nuclear accident, a new Nuclear Regulatory Commission was installed in 2012 and it was decided that it would be the head of the Nuclear Regulatory Commission who would judge on expert and technical matters relating to onsite safety at the time of a nuclear disaster rather than the prime minister serving as the head of the nuclear emergency response headquarters (Nuclear Disaster Special Measures Law, Article 20, Paragraph 3). However, regarding difficult decisions that cross the onsite/offsite boundary such as judgments on withdrawal/evacuation from a nuclear power plant at a time of rising radiation, it is not necessarily clear legally who will determine this and when. It is extremely important that rules like these are decided in advance to prevent confusion from reoccurring.

No matter what safety measures are put in place in future, there is no guarantee that Japanese society will not again suffer a serious nuclear disaster such as the Fukushima Daiichi NPS accident. At such a time, is it permissible once more for us to facilely depend on the spirit of self-sacrifice of the nuclear power plant personnel trying to stick to the field with no regard for the danger like Site Superintendent Yoshida? Will this not mean a reoccurrence of the confusion between the government and the power company on judging about withdrawal/evacuation? It is impermissible that once again who will pay what sacrifice to deal with a severe nuclear disaster is left up to the impromptu decisions of the prime minister and site superintendent as individuals.

Japanese society with its nuclear power plants must not turn its eyes away from the difficult choices imposed by worst-case scenarios in a nuclear disaster. The most crucial thing as a society as a whole is to engage in broad-ranging discussions that face head on the costs and responsibilities involved in such a scenario and to lay down convincing rules in ordinary times.

---

[1] July 22, 2011
[2] August 8, 2011
[3] An Examination of the TEPCO Videoconferences (Asahi Shuppan)
[4] July 29, 2011
[5] July 29, 2011
[6] August 8, 2011
[7] July 29, 2011
[8] July 29, 2011

What Was the Role of the Independent Investigation?

Kazuto SUZUKI

With the establishment of various investigations in response to the accident at the Fukushima Daiichi Nuclear Power Station, including the Government Investigation, the Diet Investigation and the TEPCO Investigation, what significance did the Independent Investigation have with its stance of pursuing an autonomous investigation without any organizational ties?

For one, it can be said that there were both strengths and weaknesses in being independent. Its strength was that by not belonging to any organization, it was able to maintain its neutral stance and avoid its activities being perceived in a certain light in the confusion that ensued the accident, especially when the values of opposing, maintaining, or promoting nuclear power were being strongly questioned. In contrast, however, its weakness was its independent position and because it had no ability to force compliance, it was unable to obtain information from some sources, especially the cooperation of Tokyo Electric Power. For this reason, it was unable to pursue a verification of the accident from a technical perspective.

In another sense, however, this weakness clarified the role of the Independent Investigation. As other investigations pursued technical issues as a central theme, the Independent Investigation was able to delve incisively into themes that the other investigations did not necessarily concentrate on, especially the response at the Kantei (Part 2), historical and structural issues in nuclear safety regulations (Part 3), and external relations including US-Japan relations (Part 4).

Also, because the Independent Investigation Report was the first to be published with the exception of the interim report of the Government Investigation, points discussed by the Independent Investigation set the agenda for verification of the accident by the media, in other words, it highlighted the points that needed to be examined. This paved the way for the broader investigation of social, political and cultural factors in what may have otherwise been restricted to merely a direct analysis of the accident in technical terms.

Although three years have passed since the Independent Investigation Report was published, I would like to look back to see what has been the impact of its role on subsequent issues of nuclear policy and nuclear safety regulations, and also based on the recently released Yoshida Hearings, to see to what extent the Independent Investigation was able to advance discussions even in the absence of information from TEPCO related parties, especially without any information from Site Superintendent Yoshida.

I would also like to note that this chapter is a discussion about Part 3 of the Independent Investigation Report, Analysis of Historical and Structural Factors, of which the author was in charge.

Has the Nuclear Village been dismantled?

The Nuclear Village was one of the important themes discussed in the historical and structural factors of Part 3. The "Nuclear Village" was a concept that was on everyone's lips immediately after the accident not only in the media but in the population at large, but was a concept that could give rise to misunderstanding as it was used in different ways by different people. The Independent Investigation Report pursued its analysis on "two Nuclear Villages": the Nuclear Village as a mechanism for promoting nuclear power in central government; and the Nuclear Village as a mechanism for promoting nuclear power and related facilities in local areas.

Regarding the Nuclear Village in the center, after considerable delay aided by confusion during the Democratic Party of Japan's administration, a highly autonomous Nuclear Regulatory Commission

Part 2   Three Years on from Publishing The Independent Investigation Report

was established, resolving hitherto nuclear safety regulatory governance issues such as the division of duties between the Nuclear Safety Commission and the Nuclear and Industrial Safety Agency. Also, by consciously taking a distance from industry and academia, the Regulatory Commission, its efforts to instill checks and balances in decision-making can be seen. In addition, the situation where nuclear engineers and power companies shared the same interest no longer holds, and decisions have been made that have imparted a body blow to the Nuclear Village such as thorough investigations into decommissioning and active seismic faults for aging reactors. However, points such as power consumers bearing the cost of decommissioning may well be perceived as the members of the central Nuclear Village still protecting each other.

However, the real problem is most likely the actual situation of regional Nuclear Villages. As also pointed out in the Independent Investigation Report, the financial benefits such as subsidies and property tax that go to support the regional Nuclear Villages have survived without change from before the accident. In addition, with the cessation of nuclear power plant operations one by one after the accident, local economies premised on employment and nuclear plant workers in municipalities with nuclear power plants have gone into decline and been very hard-hit coupled with an increasingly serious drop in the birthrate and an aging population. As such, in addition to still bearing the traditional diametrically opposed pattern of those for and against nuclear power, regional Nuclear Villages have started making strong demands for the restart of nuclear power plants. Moreover, given that any restart would be based on new safety standards that are the "world's most stringent", it is interesting to note that these demands now take the shape of "We want them to restart, but we want you to guarantee that they are safe at the same time."[1]

Bearing in mind these points, there would appear to be a new type of regional Nuclear Village where, instead of the old celebration of "dreamlike energy" and power-utility sponsored Mardi Gras, you find reactivating the local economy through a restart with a government assurance of safety.

Have the safety regulations changed?

In response to the accident at Fukushima Daiichi NPS, the Nuclear Regulatory Commission was founded and it was decided to set considerably stringent safety standards. Strict standards were also decided upon in the NRC's Study Team on Regulatory Standards relating to the Earthquake and Tsunamis for Light Water Nuclear Power Reactor Facilities for tsunamis, which up until then had only been viewed as a "concomitant phenomena" of earthquakes, with new standards also established for total power loss and core damage countermeasures.[2] However, can they really be said to be the "world's most stringent standards"?

There is no doubt that preparations for the safety of the nuclear power plant itself as well as for events previously deemed "unanticipated" have been greatly improved as compared to the past. Decisions on restarts based on conformity assessments also seem to be judged according to fairly in-depth evaluations. In addition, some activities such as the fracture zone survey of active faults conducted by Kunihiko Shimazaki during his tenure as a Commission member have even been deemed by some to be excessively severe.

However, the question I would like to raise here is how far the concept of "defense in depth" pointed out by the Independent Investigation has been pursued. Defense in depth refers to an idea put forward by the International Atomic Energy Agency (IAEA) in their guidelines on nuclear safety, which is an approach of putting in place multiple layers and multiple tiers of safety measures so that even if by any chance several measures are breached, the safety of the whole is ensured. The IAEA divides safety measures into five layers: 1) preventing a departure from normal operations (natural disasters and anti-terrorism, etc.); 2) preventing the expansion of an abnormality and moving toward an accident (scrum and isolation valve closure, etc.); 3) accident mitigation and preventing the abnormal release of radioactive material into the vicinity (emergency core cooling system, etc.); 4) supplementary means for severe accident response and accident management (Venting and reactor water injection from outside, etc.); and 5) disaster (inside evacuation offsite and administering iodine, etc.), the larger the scale of the accident, the deeper the layer at which action is required. Of these, measures up to the fourth layer are believed to have been put in place to a considerable extent, the problem being the fifth layer which centers on local governments and municipalities.

The Independent Investigation Report pointed out the miscomprehension of defense in depth that if you addressed a shallow layer, preparing for deeper layers was not necessary, that is, if safety measures were taken properly, there was no need to be assiduous about Level 4 severe accident measures and Level 5 disaster prevention. This was because even taking disaster prevention measures for a one-in-million "emergency" would make local residents anxious as well as to avoid friction with regional Nuclear Villages given the creation of a "myth of absolute safety."

Has this situation really changed since the Fukushima Daiichi nuclear accident and the establishment of the Nuclear Regulatory Commission? True, the development of evacuation plans only required of municipalities within a 10-kilometer radius of nuclear power plants has been expanded to 30 kilometers, involving many more municipalities in dealing with nuclear accidents. At present, it is the state that issues instructions for disaster prevention measures and local governments develop the evacuation plans, but there are no standards or assessment to determine whether these plans are effective or not, and it is hard to say that realistic evacuation plans are well established owing to various issues including financial constraints and the need to coordinate with residents.[3]

Even at the time of the Fukushima Daiichi nuclear accident, resident evacuation became a major issue, including the question of releasing SPEEDI data. With respect to the restart of nuclear power plants, developing evacuation plans for residents is included in the conformity assessment, but if you were to question whether they were really examined properly for effectiveness, satisfactory assessments have not necessarily been carried out, as can be seen by the approval of a restart for the Sendai Nuclear Power Station,.

Has the "Safety Myth" collapsed?

The underpinning concept used throughout the Independent Investigation Report, especially Part 3, was the idea that "a logic could be formulated in the direction of an ultimate premise that [the safety of nuclear power] was not to be doubted and that 'safety' could be determined in advance"[4], in what was said to be the "nuclear power safety myth" or the "myth of absolute safety". Chapter 9, Part 3 of the Independent Investigation Report showed that the "safety myth" was essential in the process of introducing nuclear power generation to a Japan that had suffered atomic bomb strikes, but has this "safety myth" really collapsed in the wake of the Fukushima Daiichi nuclear accident?

The collapse of the "safety myth" had also formerly been advocated after the Three Mile Island and Chernobyl accidents, but each time the discourse "Accidents that occur overseas won't happen in Japan. Japanese nuclear power plants are safe" was used to enhance the myth, failing to lead to a substantial collapse thereof. This was because anxiety over radiation that could not be seen by the naked eye and possible future accidents was always present and a "safety myth" was necessary in order to dispel such anxiety and keep the nuclear power plants running. However, the myth that "Japan's nuclear power plants are safe" is no longer tenable after seeing those fears realized, the hydrogen explosion in the reactor building, the forced evacuation of many people, and the destruction of their daily lives.

However, the fact that there were no direct deaths from radiation caused by the nuclear accident (but the number of nuclear-related deaths in the course of the evacuation was over 1000), that the accident did not result in a worst-case scenario, that designated evacuation areas have gradually decreased through decontamination, meant that the fears anticipated immediately after the accident did not eventuate, leading to diverse social reactions to the accident by the people. On the one hand, there were those who emphasized the dangers of nuclear power plants and called for the immediate closure of all nuclear power plants, while on the other hand, there were those who worried about the impact on electricity prices being raised in a serious of stages and economic activity and who called for the nuclear power plants to be reopened. This diversity of values has made it difficult for the "safety myth" to be revived.

Four years, however, have passed since the nuclear accident and the anti-nuclear demonstrations in front of the Kantei that initially exceeded 10,000 people have gradually subsided, and as the voice of people emphasizing the damage from radiation grows smaller, the reality is that there is a growing mood in favor of reopening the nuclear power plants. Although there was violent opposition when Ooi Nuclear Power Station was reopened due to power shortages at Kansai Electric Power, the same extent of opposition was not seen when Sendai Nuclear Power Station passed the conformity assessment and obtained the agreement of local residents. This quiet and gradual toleration of plant reopening may well be supported by new safety myths.

The first of these new safety myths is the myth that "they're safe because the state has examined them by the world's most stringent standards." As already mentioned, despite insufficient disaster prevention measures at the fifth layer of defense in depth, this can be attributed to a judgment that it is still safe because the state passed them. Of course, only experts can determine the safety of nuclear power plants. However, as long as the perception that it is "safe" because the experts say so exists, very little difference with the former "safety myth" exists. There is the possibility of an accident occurring if power plants are reactivated and in order to prevent accidents, measures need to be put in place, including the fifth layer of defense in depth, but even after the Fukushima nuclear accident, since "misunderstandings over defense in depth" remain unresolved, there are signs of revival in the "safety myth".

Another new safety myth can be said to reflect the hopes of

regional Nuclear Villages. Although this existed even prior to the Fukushima Daiichi nuclear accident, nuclear power plants are an unrivaled economic resource for the regions and their reopening is positioned as a key issue in recovering economic benefits lost by the nuclear power plant shutdowns. As a result, they need to go ahead with reopening after dispelling local fears. For example, at the time of the reactivation of the Sendai NPS, Yuichiro Ito, the Governor of Kagoshima Prefecture, made statements such as "no life-threatening problems will happen"; "It's the view of the state that for the time being we don't have any choice but to take advantage of nuclear power"; "Even if an accident happens, radiation 5.5 kilometers from the plant will be 5 microsieverts/hour (…) no evacuation is needed. You can live normally."[5] Here it is seen that the measures are unavoidable for economic reasons, that there is a promise of safety from the Nuclear Regulatory Commission, and that a worst-case scenario has been excluded by the experiences of the Fukushima Daiichi nuclear accident.

Although these new safety myths are not the blind mythology of former times, they can be considered as comprising elements such as economic grounds, guarantees by the state, and an abuse of experiences from the Fukushima Daiichi nuclear accident.

The submerged generator accident pointed out in the Yoshida Hearings

It was difficult in the research conducted by the Independent Investigation to interview TEPCO parties and to access data from the site that had not been publically released, which was an undeniable weakness of the report as already mentioned. The public release of the various interviews conducted by the Government Investigation, however, has made it possible to compare head on the Independent Investigation's report with a variety of hearings, including the Yoshida Hearings. While it is not possible to verify every topic given the enormous number of working papers that have been released, I would like to touch on several points that were not discussed fully in the Independent Investigation.

Part 2 Three Years on from Publishing The Independent Investigation Report

One is that we were not able to adequately cover the emergency diesel generator (D/G) submersion accident at the Fukushima Daiichi NPS of 1991. In October 1991, the emergency D/G in the basement was submerged due to a pipe leakage, triggering an accident that put the emergency D/G out of action. The accident was announced by the Agency for Natural Resources and Energy on October 31 of the same year[6] with drawings also made available on the following December 20, but the Independent Investigation was unable to mention this issue. NHK also picked up the news of this accident on December 29, 2011, but here it was reported as a report heard from a "former TEPCO employee".[7]

In the Yoshida Hearings, it was understood from the experience of this submersion accident in 1991 that it would take time for recovery in the case of the emergency D/G being immersed in water. It was recorded that at the time the submerged D/G was dismantled and dried out and inspected over a period of some six months, but from the experience of this accident, the idea of avoiding submersion by installing the emergency D/G in a higher place, for example, was not reflected in Site Superintendent Yoshida's account.

In fact, looking at the Agency for Natural Resources and Energy's documents, the measures taken were "in addition to replacing the component cooling water system water piping, they decided to implement theoretical seawater pipes in order to improve verifiability and maintainability and so on. Note that it has been decided for the replacement to use material that has excellent adhesion to the pipe inner surface and peeling resistance in comparison to the current pipe lining"[8], the main focus being to improve the tubing.

Even if the emergency D/G had been reinstalled in a high place, they may still not have been able to avoid a total power outage (SBO) because as long as the switchboard was in the basement, it would be unable to receive the power supply from the emergency D/G if it was swamped by water from a tsunami. However, despite the fact that the emergency D/G, which could be deemed to be the lifeline for ensuring nuclear power plant safety, had been submerged in the past, there is no evidence that the Nuclear Safety Commission, the Agency for Natural Resources and Energy or NISA discussed how to deal with this. I believe the fact that it is mentioned even in Yoshida Hearings merely as a referent for restoring the emergency D/G is an issue that should have been discussed from the point of view of nuclear safety regulations.

---

[1] For example, Asahi Shimbun, Kagoshima Governor 'State guarantees safety' Regarding restart of Sendai NPS, June 13, 2014

[2] An outline of the new standards is provided in Nuclear Regulation Commission, New Regulatory Standards on Commercial Power Reactors and Nuclear Fuel Facilities, no date;
https://www.nsr.go.jp/activity/data/20140214.pdf.

[3] The NHK program Contemporary Close-Up ran a feature on this, highlighting the problem. NHK, The Nuclear Accident: How to Protect Residents' Safety, Contemporary Close-Up, August 27, 2014.
http://www.nhk.or.jp/gendai/kiroku/detail02_3541_all.html

[4] Independent Investigation Report, Section 1, Chapter 8, Part 3

[5] All citations from a Mainichi newspaper article. Mainichi Shimbun, Kagoshima Sendai NPS: restart agreement", November 8, 2014.
http://mainichi.jp/shimen/news/20141108ddm041040134000c.html

[6] Agency for Natural Resources and Energy, Press Release: Manual Stop at Unit 1 of the Tokyo Electric Power Co., Ltd. Fukushima Daiichi Nuclear Power Station, October 31, 1991.
http://www.nsr.go.jp/archive/jnes/atomdb/events-data/events-000326.html

[7] NHK, TEPCO emergency power submerged in the past, December 29, 2011.
http://www3.nhk.or.jp/news/genpatsu-fukushima/20111229/0445_suibotsu.html

[8] Agency for Natural Resources and Energy, "References: Causes and Measures for Manual Stop at Unit 1 of the Tokyo Electric Power Co., Ltd. Fukushima Daiichi Nuclear Power Station, no date.
http://www.nsr.go.jp/archive/jnes/atomdb/events-data/events-000323.html

Part 2   Three Years on from Publishing The Independent Investigation Report

My Thoughts in Light of the Release of the Yoshida Hearings

Shinetsu SUGAWARA

In the four years since the Fukushima nuclear accident, valuable information that could not be accessed at the time by the Independent Investigation is becoming increasingly clear. While there is much room for discussion about the circumstances leading up to the public release of the Yoshida Hearings (and other records of the Government Investigation hearings), as a member of the nuclear power community, I was struck upon reading its contents by just how valuable a record it is, providing very many pointers.

In Part 3 of the Independent Investigation, we examined the historical and structural factors why "preparations" for the risks associated with the use of nuclear power were insufficient. However, since we were unable to conduct adequate interviews with the nuclear power business (TEPCO), which by right bears the primary responsibility for nuclear safety, the discussion per force had to focus on safety regulations and the relationship with local communities. Moreover, due to various constraints, we were unable to adequately provide a clear direction on how to resolve and/or improve the structural issues that were identified.

Based on the variety of efforts concerning nuclear power that have taken place since the accident (the review of nuclear safety regulations and the voluntary efforts of nuclear operators etc.), I would like to point out two points below gleaned from the many published working papers that I believe to be important.

The need for risk management and regulatory science

"In a sense, though it's funny to call it irresponsible,

> *academics can say there's a possibility as much as they like, but when you ask if it's at a properly designable level or something, it isn't."*
>
> (Yoshida Hearings, November 6, 2011)
>
> *"Now, after having run into serious trouble, I think it would've been better to have done something, but that's hindsight, so if you think about conditions before the earthquake came, you can only respond when an established theory emerges and academia and the experts provide a proper direction."*
>
> (ibid.)

As has already been pointed out many times, knowledge on the Jogan Tsunami (869) was gradually being accumulated from several years before the accident. However, the Great East Japan Earthquake hit without those findings being utilized in NISA's regulations or TEPCO's tsunami measures. Since the Independent Investigation was unable to obtain first-source information about TEPCO's past decisions, this resulted in a preoccupation from start to finish with how tsunami risk was handled in safety regulations. However, the fact that TEPCO knew about pioneering research on the Jogan Tsunami and still put off the consideration of specific tsunami measures for the reasons that "the scientific knowledge is still not solid" and "it hasn't been authorized by the Society of Civil Engineers, the authority in this field" is vividly recounted in the Yoshida Hearings.

In risk management in fields which closely involve science and technology, the general process is to determine the pros and cons of measures based on a scientific "risk assessment" while taking into account other social and economic factors, in other words to make decisions on "risk management". If the above response to the Jogan Tsunami is considered in this context, it amounts to a schema of postponing decision-making for "risk assessment" (=specific tsunami measures) because of substantial uncertainty (=scientific knowledge of the tsunami). Of course, it is also possible that sufficiently based on risk assessment, risk

Part 2 Three Years on from Publishing The Independent Investigation Report

management decisions may be "to take no measures". However, given the current accident, the decision to "not do anything for the time being" because of premature scientific findings should be considered as unacceptable in the future, at least in the nuclear field.

On the other hand, Site Superintendent Yoshida has harshly criticized the regulatory authorities at the time, "NISA was basically a government institution that did nothing itself but just hid from taking responsibility behind a shield of academic opinion and so on" (November 6). Taking a step back, however, the same criticism holds true for the operators' own behavior, which was to postpone tsunami measures on the grounds that the Society of Civil Engineers' opinion was not solid, thereby "hiding from taking responsibility" for risk management because of "academic opinion".

What is required today in the wake of the Fukushima accident at even a stage encompassing substantial uncertainty in risk assessment would be how to derive as rational (or least regrettable) risk management decisions as possible. This is close to the concept known as "regulatory science" that is emphasized in the field of risk management of chemical substances and food safety. For example, in the risk management of chemical substances, the process adopted is to determine specific risk management measures (regulation standard values, etc.) based on risk assessment results derived from the relevant natural sciences (such as toxicology, immunology), accounting for various uncertainties (such as estimates of the impact on human health based on animal experiments, individual human differentials) by multiplication by a constant factor. Certain "protocols", "conventions" and knowledge on the state of systemic design that supports them is gradually being accumulated to enable "not-so-off-the-mark" risk management decision-making even when risk assessment itself includes uncertainties and the unknown.

It is a fact that this idea of determining reference values, etc. while assuming the uncertainty of scientific knowledge has always been carried out both implicitly and explicitly, even in engineering fields like nuclear power. Engineering is the business of grasping social needs, placing them into specific boundary conditions, then designing actual equipment and systems. It is conventional in this process to "draw the line" on boundary conditions assuming a suitable safety factor, and even in the safety regulations, the practice has been to decide reasonably "conservatively" (on the safe side). However, as the observation that the "conservative" deterministic approach that has been used in conventional safety regulations has become "an opaque approach where experts monopolize conservative judgments" (Government Investigation hearing with Kondo Shunsuke, NRC Chairman) shows, to "draw the line" like this by experts may draw criticism that it is fraught with arbitrariness and intransparency. Moreover, when this drawing the line is in fact carried out by a very limited number of stakeholders such as design officers in the field as in the past, there is a possibility that related knowledge existing in various places throughout society is not reflected sufficiently. In addition, the point that since value judgments over "safety" levels and methods for their achievement are inevitably involved in "drawing the line", if the interests of the community known as the Nuclear Village are posited, various political and social relations may distort those value judgments of the experts is one mentioned by the Independent Investigation at great length.

So, in particular regarding issues involving large risks and uncertainties, rather than have this "drawing the line" decided only by limited experts, greater sophistication is required for a risk management "protocol" that "decides appropriately despite uncertainties" all while gaining input from various stakeholders. I consider making convincing judgments that are fair to a certain degree from a variety of stances for what has conventionally been decided by expert judgment, as well as linking this to a realistic mitigation of risk, should be the essence of future "regulatory science" in the field of nuclear power.

In Japan, when uncertainty of risk assessment is large - foregoing the consideration of measures because scientific knowledge is insufficient until knowledge matures (for example, responses

Part 2 Three Years on from Publishing The Independent Investigation Report

to the Jogan Tsunami), or immediately banning certain activities because scientific knowledge is insufficient (for example, regulatory response on the fracture zone in the Japan Atomic Power Company's Tsuruga Nuclear Power Station) and so on - there are not a few cases where responses to uncertainties oscillate wildly. While it is of course possible that taking into consideration various factors can lead to decision-making to "do nothing" or "ban immediately", there is a broad scope of options that can be taken as risk measures for complex systems and manmade objects, and it is my hope that we can become more socially proficient at handling risks using the "regulatory science" approach as a point of reference.

Former Japan Atomic Energy Commission Chairman Kondo's "remorse"

> *My biggest mistake was to choose the occurrence probability of excessive exposure to the public as the index in the discussion of safety goals. I was convinced that protecting lives was safety. (...) I should have chosen the occurrence probability of land contamination beyond the site boundary based on level 3PSA[1]. It was a real blunder not to have learnt from the Chernobyl accident the lessons about the depth of community collapse as well as where to set up an offsite center.*
>
> (Government Investigation's Kondo hearing)

Of all the series of hearings released, it is the above remorse expressed by former Chairman Kondo of the Japan Atomic Energy Commission, who led the field of nuclear safety in Japan, that leaves a strikingly strong impression on me. It demonstrates his efforts to provide a quantitative answer to the fundamental question surrounding nuclear safety of "how safe is safe enough?" and is a very important feature as a yardstick of risk management. The Independent Investigation touched on the fact that the development of safety goals for nuclear power was behind in Japan, and that it had not been clearly positioned in the regulatory system but limited to the interim report of the Nuclear Safety Commission, pointing out that the job of dealing properly with risk in safety regulations was backward even compared to other countries.

The Nuclear Regulatory Commission, which was newly established after the accident, "determined" its safety goals in April 2013. Taking into consideration environmental contamination by radioactive substances, these include the phrase "the frequency of occurrence of an accident where the cesium (Cs) 137 discharge exceeds 100 terabequerels (TBq) should be restrained so as not to exceed once in one million reactor years. In this sense, this can be seen as some kind of allowance having been made for former Chairman Kondo's "remorse". However, the development process was no more than the Commission discussing several times on the basis of the former NISA's interim report and unfortunately was not based on a widespread social debate.

Since safety goals are an index of what level of risk ("safety" level) is socially permissible as described above, they need to be decided not purely technically but in light of social discussion. From the viewpoint also of reducing as much as possible any situation that would cause the kind of regret by experts that the Kondo hearing demonstrates, I believe it would be beneficial to obtain a variety of inputs in the development process through dialogue with diverse stakeholders including experts in surrounding fields, local governments and residents. In particular, nuclear power has been a field with a strongly comprehensive engineering nature from its outset, and in the case of a technology that encompasses huge hazards and once an accident occurs, has a massive impact on diverse aspects of the social economy, this makes the need to involve diverse stakeholders exceptionally strong. Furthermore, safety goal setting itself is of course important, but deepening the debate at a societal level over "risk" and "uncertainty" is also important in the process of determining safety goals, which would also contribute to "improving the risk literacy of society as a whole" so often emphasized nowadays.

Part 2   Three Years on from Publishing The Independent Investigation Report

I believe the Independent Investigation's mention of safety goals was significant, but in the light of ensuing developments, it is a shame that it did not go further in pursuing the social significance and the development process.

Elsewhere, the nuclear power operators appear currently to be somewhat reluctant about this kind of discussion owing to the fact that the development of safety goals by the regulatory authorities was not made in a very clear form. It is considered essential that operators set risk management objectives in terms of managing risk voluntarily, and setting goals, for example, while discussing with local communities surrounding nuclear facilities is also important in the practice of "risk communication".

After joining the Independent Investigation, the author was admitted to the Central Research Institute of Electric Power, choosing to pursue a career at a research institute in the power and nuclear industry. From October 2014, he has been serving concurrently at the newly established Nuclear Risk Research Center in-house, literally working on nuclear risk management as an important proposition. Therefore, the issues noted here are by no means directed at others, but are seen as part of his own brief. Bolstered by his time at the incomplete but rare and stimulating venue experienced at the Independent Investigation, he wishes to continue working in the future on these difficult challenges.

---

[1] Probabilistic Safety Assessment (PSA) is an established technique to numerically quantify risk measures in nuclear power plants.
   Level 1 PSA: The assessment of plant failures leading to the determination of core damage frequency.
   Level 2 PSA: The assessment of containment system and phenomenological responses, leading to the determination of containment release frequencies.
   Level 3 PSA: The assessment of the off-site consequences, leading to estimates of adverse impact on the health of the public

Afterword

After reading the Hearings, a "lieutenant" of Site Superintendent Masao Yoshida's responding around the clock to the accident at the Roundtable in the Emergency Response Center at Fukushima Daiichi Nuclear Power Station told me, "I was impressed that Yoshida-san could remember so much."

"Yoshida-san says in the Hearings that 'there are gaps in my memory, I'm not sure of the order of things', and I also agree. However, Yoshida-san remembered things well. My recollection is clear up until the explosion in Unit 1, but it's ambiguous after that…"

Since the hearings took place more than six months later, it was only natural that the memory would become blurred about the sequence of events in mid-March of 2011. Yoshida possessed the tenacious will vital to a commander in a time of crisis, but he also possessed a tenacious memory.

However, the Hearings only took the form of answers to questions from the Government Investigation officers. Yoshida did not answer basically what was not asked. And, there were many questions the Government Investigation officers did not ask.

Take, for example, the following points.

"At the time (…) I also heard talk that CEO Shimizu had said to Kan 'Let us withdraw'. I don't know whether that's linked to something I said to someone at Head Office being passed on to Shimizu or what I said to Hosono-san, but there was some kind of double-line talk." (Yoshida Hearings, August 9)

When and from whom did Yoshida hear "talk that CEO Shimizu had said to Kan 'Let us withdraw'"?

Another one of the TEPCO engineers holed up at the Roundtable at the time told me, "No such talk was heard at the Roundtable around that time. Yoshida-san probably heard it later" and at the same time stated, "sensitive subjects like evacuating or withdrawing weren't dealt with by videoconference, but through direct interaction with the parties concerned by Yoshida over the phone", so the truth remains cloudy.

So, what did Yoshida feel when he heard that story? Did he ask someone at Head Office? Did he question Shimizu directly? How, in the first place, did the "double-line" become tangled over the matter of "withdrawal"?

On this day, the night of March 14, what kind of countermeasures for the "withdrawal" were discussed by the management team at TEPCO Head Office? Was Yoshida asked anything at the time?

No follow-up questions like these are made.

Furthermore, the Government Investigation hearings were not conducted as a series of conversations or "chats" to reawaken the memory in a natural way, but rather to verify predetermined assumptions on the facts. No doubt this approach of focusing on certain (judgments and responses perceived as) decisive errors and ascertaining the facts thereof was the most effective way of investigating the cause of the accident and its expansion in a limited timeframe.

However, the use of this approach alone makes it easy to miss the eruption of feelings and magma of thoughts underpinning the decision-making.

Nevertheless, comparing the questions of August and November, those of November are more sophisticated in terms of content and the questioning more probing. I am not alone in feeling a certain exhilaration as the truth became clearer. A case in point is the questions about preparations for an earthquake and tsunami, which provide a wealth of hints in terms of lessons from the accident. Yoshida's answers to these show his lingering sense of heartfelt thoughts and pathos. This is also one of the results of the November hearing.

I previously wrote that this record was a "testimony", but it is only when we who remain behind extract the lessons learned from it that the "testimony" takes on a historical meaning. An executive at TEPCO Head Office said to me "[the Hearings] are like a department store of essential lessons."

It being a matter of course that TEPCO must learn the lessons it imparts, many other challenges abound for us concerning a culture of nuclear safety, especially with regard to nuclear power plant severe accident response.

And to reiterate, a tireless investigation of the truth is essential for learning those lessons. It is important that repeated verification takes place and feeds back into actual actions.

I spoke to five TEPCO employees while deciphering the Yoshida Hearings. Three of them were involved with Site Superintendent Yoshida in the onsite response in the midst of the crisis. Another worked as an assistant to top management at Head Office at the time, and the last one is now responsible as a member of the management team for crisis management and embedding a culture of nuclear safety. I was asked not to release their names due to their respective positions, but I would like to thank them all for according me an interview. In addition, Professor Akira Omoto of the Tokyo Institute of Technology, who served on the Atomic Energy Commission, was kind enough to respond by e-mail to my inquiry about the contents of the Yoshida Hearings. I would like to express my deep thanks to him.

The Rebuild Japan Initiative's attempt to unravel the Yoshida Hearings by members of the Independent Investigation's working group, and in the light of that to examine the Independent Investigation Report, as well as co-conducting a symposium with the Keio University Global Security Research Institute

(GSEC) – The Fukushima Crisis As Seen From the Yoshida Hearings – and even publishing this report, have all been conducted in the hope that it should become part of constructing this kind of feedback mechanism.

Toshihiro Okuyama of the Asahi Shimbun Editorial Board provided very appropriate guidance as our Advisor.

I asked Takashi Otsuka, who was kind enough to act as editor for the Independent Investigation Report, to once again be editor for this report. As before, it was a rush job with a last-minute deadline, but his editing was consummate.

As with the Independent Investigation Report, Kay Kitazawa (Research Director, Rebuild Japan Initiative) once again acted as staff director for the project. Under her guidance, Natsuki Fujita served as intern under Kitazawa, doing an impressively solid job.

And finally, I would like to dedicate this work to Koichi Kitazawa, a former president of the Tokyo City University (and former chair of the Science and Technology Agency), who passed away last September.

Professor Kitazawa served as Chairman of the Independent Investigation. Without his sense of mission and responsibility as a scientist there would have been no Independent Investigation and no Report.

After the publication of the Independent Investigation Report, Professor Kitazawa maintained a keen interest in the investigation of the causes of and background to the Fukushima nuclear accident, and also participated in the Rebuild Japan Initiative's Japan's Lost Era investigation project, authoring the chapter "The Fukushima nuclear accident: Lost Opportunities and the "safety myth". It was his last paper. This is to be published under the title of "Examining Japan's Lost Decades" in both Japanese (Toyo Keizai Inc., May 2015) and English (Routledge, 2015).

Had he still been alive, I am sure he would have agreed with this project.

I remember fondly his warm but at times stern face.

February, 2015
Yoichi FUNABASHI
(Chairman, Rebuild Japan Initiative Foundation)

PROJECT MEMBERS

Yoichi FUNABASHI is Chairman of the Rebuild Japan Initiative Foundation. He is a graduate of the Faculty of Education, Tokyo University. He joined the Asahi Shimbun in 1968. After being a Nieman Fellow at Harvard, he served as correspondent for the Asahi Shimbun in Beijing and Washington, and American General Bureau Chief, later serving as Editor-in-Chief for the Asahi Shimbun until October 2007. He took up his current post in September 2011. He acted as the Project Director for the Independent Investigation Commission on the Fukushima Nuclear Accident.

Kazuto SUZUKI is a Professor at the Graduate School of Law, Hokkaido University (member, Panel of Experts, UN Security Council Iran Sanctions Committee) He specializes in international political economy and EU research. In 2012, he received the 34th Suntory Prize in Space Development and International Politics. He served as leader of the Working Group for Part 3 of the Independent Investigation.

Shinetsu SUGAWARA is a Senior Researcher at the Institute of Social and Economic Research, General Central Research Institute of Electric Power Industry. He specializes and is interested in nuclear power policy and the social theory of science and technology. He graduated from the doctoral course of The University of Tokyo Graduate School of Nuclear Engineering, International Nuclear Program. In April 2012, he joined the Central Research Institute of Electric Power, serving concurrently in that Institute's Nuclear Risk Research Center from October 2014. He served as a member of the Working Group for Parts 1 and 3 of the Independent Investigation.

Akihisa SHIOZAKI is a lawyer dealing mainly with corporate governance and risk management. He was secretary to the Chief Cabinet Secretary in 2006-07. He is Vice Chair of the Civil Intervention Violence Committee, the Dai-ichi Tokyo Bar Association. He is auditor of the Rebuild Japan Initiative and served as a member of the Working Group for Part 2 of the Independent Investigation.

Kenta HORIO is an expert investigator in the Japanese mission of an international organization in Vienna. He specializes in nuclear energy policy and nuclear non-proliferation. He has been enrolled in the doctoral course at The University of Tokyo Graduate School of Engineering, International Nuclear Program since April 2010. He assumed his current position in May 2013. He served as a member of the Working Group for Part 1 of the Independent Investigation.

Takashi OTSUKA is a science journalist and served as the Project Editor. He is a graduate of the Faculty of Engineering, Kyoto University. He joined the Asahi Shimbun in 1976, serving as staff science correspondent in the American General Bureau 1992-95, and Science News Editor at the Tokyo head office 2001-2004. He was President & CEO of Asahi Shimbun America 2004-2006. He also served as editor for the Independent Investigation Report.

Kay KITAZAWA is Research Director at the Rebuild Japan Initiative Foundation. She graduated from the Faculty of Arts, The University of Tokyo. She is also a graduate of the Graduate School of Frontier Sciences at the same university. She is involved in Geographic Information Systems (GIS) consulting mainly in Europe as part of her PhD study at The Centre for Advanced Spatial Analysis, University College London. She is in charge of research consulting on urban policy private sector think tanks. She served as the staff director for the Independent Investigation.

Toshihiro OKUYAMA served as the Project Advisor. He is a member of the Asahi Shimbun editorial board and graduated from the Department of Nuclear Engineering, Faculty of Engineering, The University of Tokyo. He joined the Asahi Shimbun in 1989 and has served as a journalist in the Mito Branch, Fukushima Branch, Social Affairs Department, and Special News Department among others.

Natsuki FUJITA was an intern for the project. He is a graduate of the Faculty of Policy Management, Keio University. He also served as an intern in the Independent Investigation.

Fukushima Nuclear Accident Chronology of Events

TIME	DISASTER/ ACCIDENT	FUKUSHIMA DAIICHI	TEPCO	METI/NISA	PM/KANTEI
3/11 14:46	Great East Japan Earthquake	Emergency shutdown at Fukushima Daiichi, Fukushima Daini (14:46) Unit 1 IC starts (14:52)			PM instructs establishment of Earthquake Response Center at Kantei
15:00	First tsunami (4 meters) hits Fukushima Daiichi (15:27) **Second tsunami (14–15 meters) hits Fukushima Daiichi (15:35)** **Unit 2 total power outage (15:41)** Unit 1 water indicator out of action (15:51)	Manual start of RCIC in Unit 2 (15:02) Manual shutdown of Unit 1 IC (15:03) Manual start of RCIC in Unit 3 (15:05) Unit 3 RCIC stops (15:25) Unit 2 RCIC stops (15:28) Unit 2 RCIC manual start (15:39)	Emergency Response Center set up at Head Office (quake damage assessment, power loss restoration measures) (15:06) Deem total power loss at Units 1-5, Fukushima Daiichi; contact NISA (15:42) (However, Units 4 & 5 report later amended) First announcement of emergency; Emergency Response HQ Integrated with Emergency Disaster Response HQ (15:42)		PM arrives at basement Crisis Management Center (around 15:00) ERC set up (15:14) First meeting of ERC (15:37)
16:00	Unit 1 core starts to be exposed (NISA analysis) (around 16:40)	Unit 3 RCIC manual start (16:03) Deem Units 1 & 2 emergency water cooling equipment out of action because unable to verify status of nuclear reactors and water pumping status unknown (16:36)	Second announcement of emergency (16:36) Notify NISA of Nuclear Disaster Law Article 15 event at Units 1 & 2 (emergency water pumping out of action) (16:45) All TEPCO's high- & low-pressure power trucks start heading to Fukushima (around 16:50)		First PM press conference (16:54)

TIME	DISASTER/ACCIDENT	FUKUSHIMA DAIICHI	TEPCO	METI/NISA	PM/KANTEI
17:00		Site Superintendent Yoshida instructs staff to start looking into pumping water into the reactors using fire trucks and fire extinguishing lines set up as an accident management measure (17:12)		METI Minister reports to PM on reactors and submits proposed Declaration of Nuclear Emergency 1 (17:42)	
18:00	Unit 1 reactor damage starts (NISA analysis) (around 18:00)	Unit 1 IC, backwash valve opening manoeuvre (18:18) Closing manoeuvre (18:25)			
19:00					State of Nuclear Disaster announced; Nuclear Disaster Response HQ set up; first meeting of Nuclear Disaster Response HQ (19:03) First Chief Cabinet Secretary press conference (19:45)
20:00	Fukushima Prefecture issues evacuation order for residents within 2km of Fukushima Daiichi NPS (20:50)				
21:00		Unit 1 IC, backwash valve opening manoeuvre (21:30)			Instructions for residents within 3km radius of Fukushima Daiichi NPS to evacuate. Indoor evacuation order for 3–10km radius (21:23)
22:00	Unit 1 containment vessel damaged (NISA analysis) (around 22:00)	Arrival of first group from Tohoku Power and one high-pressure power vehicle verified (around 22:00)			
Night		From late on the 11th to early on the 12th, power trucks from Tohoku Power and TEPCO arrive			
3/12					
0:00		Site Superintendent instructs to prepare to vent Unit 1	TEPCO notifies METI Minister of Nuclear Disaster Law Article 15 event at Units 1 & 2 (extraordinary rise in containment vessel pressure) (0:55)		
1:00				Receive SPEEDI estimates from Nuclear Safety Center (1:12) Send estimates to NISA liaison at Kantei Operations Room (around 1:35)	Kantei approves TEPCO request to vent (around 1:30)

TIME	DISASTER/ACCIDENT	FUKUSHIMA DAIICHI	TEPCO	METI/NISA	PM/KANTEI
1:00				Operations Room Director shares this material with related ministries (1:41)	Kantei ERC releases 'Future Developments in Unit 2 at Fukushima Daiichi' prepared by NISA (1:40)
3:00			METI & TEPCO press conference on vent implementation (3:06)		Chief Cabinet Secretary mentions vent in press conference (3:12)
4:00		Water pumping via fire engine extinguishing line starts; 1.3 tons injected (around 4:00)			
5:00		Restart pumping into Unit 1 using fire trucks (5:46)			Evacuation order for residents within 10km radius of Daiichi (5:44)
6:00				METI Minister orders TEPCO to vent (6:50)	PM leaves the Kantei for field visit (6:14)
8:00		Site Superintendent instructs Unit 1 vent (8:03)	Confirm partial evacuation completed in Okuma Town (8:27)	Receive report from TEPCO to the effect "Unit 1 vent to commence around 9 a.m. The water level in Unit 1 has dropped close to Top of Active Fuel, water is being pumped in via fire extinguishing pumps" (8:29)	PM departs Fukushima Daiichi (8:05)
9:00		Duty officers leave to conduct the vent (9:04) MO valve in Unit 1 reactor opened manually (around 9:15)	Confirm evacuation completed in Okuma Town (9:03)	Receive report from TEPCO "first valve is open" (9:30)	
10:00		Open AO valve in Unit 1 pressure chamber (10:17)	Since dosage is climbing, deem high possibility that radioactive material released during vent (10:40)		PM arrives at the Kantei (10:47)
11:00		Unit 3 RCIC stops (11:36)	Since dosage is falling, confirm vent may not be working sufficiently (11:15)		
12:00		Unit 3 HPCI starts automatically (12:35)			

TIME	DISASTER/ACCIDENT	FUKUSHIMA DAIICHI	TEPCO	METI/NISA	PM/KANTEI
14:00			Confirm containment vessel pressure falling, deem this due to release of radioactive material by the vent (14:30)		
		80 tons fresh water pumped into Unit 1 by fire trucks (14:53)	CEO Shimizu confirms and approves of seawater injection (14:50)		
		Site Superintendent issues instructions to implement seawater injection into Unit 1 reactor (14:54)			
15:00			Notify ministries etc. re drop in containment vessel pressure (15:18)		
	What seems to be a hydrogen explosion takes place in Unit 1 (15:36)	With the restoration of power via power trucks, reparations to inject boric acid into Unit 1 reactor completed (15:36)	Notify NISA and Cabinet Information Center of plan to inject seawater (around 15:20)		
16:00			Notify METI Minister of Nuclear Disaster Law Article 15 event at Units 1 & 2 (extra-ordinary rise in radiation at site boundary) (16:27)		
17:00		Site Superintendent issues instructions to prepare to vent at Units 2 & 3 (17:30)		METI Minister order to TEPCO to fill the reactors with seawater (17:55)	Chief Cabinet Secretary press conference: 'Some kind of explosion' (17:45)
18:00		Confirm electrical equipment and hoses for boric acid injection damaged by scattered debris and can't be used (around 18:30)			Evacuation order for residents within 20km radius of Daiichi (18:25)
19:00		**Unit 1 seawater (no boric acid) injection by fire trucks starts (19:04)**			PM gives METI Minister seawater injection order (19:55)
20:00				Document ordering seawater injection finished (20:05)	**PM Message 'I must ask you to evacuate' (20:32)**
		Start injecting mixture of seawater and boric acid into reactor vessel (20:45)			Chief Cabinet Secretary press conference 'reactor vessel did not explode' (20:41)

TIME	DISASTER/ACCIDENT	FUKUSHIMA DAIICHI	TEPCO	METI/NISA	PM/KANTEI
3/13					
2:00		Unit 3 HPCI stops (2:42)			
5:00		Deem Unit 3 to have lost cooling function (5:10) Instructs to start completing Unit 3 vent line-up (5:15)	Notify Nuclear Disaster Law Article 15 event (loss of reactor cooling function) (5:58)		
7:00	Unit 3 reactor exposure begins (NISA analysis) (around 7:40)				
8:00		Vent line completed for Unit 3 (8:41)	Notify ministries etc. Unit 3 vent line completed (8:46)		
9:00		Confirm drop in containment vessel pressure due to Unit 3 vent (around 9:20) Start injecting mixture of seawater and boric acid into Unit 3 (9:25)	Notify ministries etc. of seawater injection into Unit 3 reactor (9:36)		
10:00		**Site Superintendent issues instruction to vent at Unit 2 (10:15)** Site Superintendent issues instruction to prepare for possible seawater injection at Unit 3 (10:30)			
11:00		Vent line completed at Unit 2 (11:00)			
13:00		Start seawater injection at Unit 3 (13:12)			
19:00					PM Message (request to public re brownouts and blackouts) (19:49)
22:00	Unit 3 pressure vessel damaged (NISA analysis) (around 22:10)				

TIME	DISASTER/ACCIDENT	FUKUSHIMA DAIICHI	TEPCO	METI/NISA	PM/KANTEI
3/14					
11:00	Unit 3 Building explodes; water pumping lines also damaged at Unit 2 (11:01)	With the explosion at Unit 3, Unit 2 vent valve closes (11:01)			
13:00			Since water level falling in Unit 2, notify ministries etc. of imminent seawater injection to the reactors (13:18)		
16:00		Switch fire trucks and hoses; build pumping line and restart seawater injection (16:30)			
18:00	Unit 2 reactor exposure begins (NISA analysis) (around 18:00)	Water level in Unit 2 reactor drops; deem all fuel exposed (18:22)			
19:00	Unit 2 reactor damage starts (NISA analysis) (around 19:50)	Start pumping seawater by fire trucks into Unit 2 reactor (19:54)	Notify ministries etc. of Unit 2 total fuel exposure (19:32)		
20:00			VP Sakae Mutoh gives press conference on Unit 2 reactor (20:40)		
21:00		Unit 2 vent line completed (21:00)			
22:00	Unit 2 pressure vessel damaged (NISA analysis) (around 22:50)				
3/15					
0:00		Unit 2 Dry Well (D/W) valve opening operation; vent line completed (0:02) A few minutes later, confirm valve closed			
3:00		Since Unit 2 containment vessel pressure surpassed design limits, confirm state of insufficient depressurization (3:00)		Discussion between PM, METI Minister, Matsumoto Minister for the Environment, 3 deputy secretaries, 3 assistant secretaries, Nuclear Safety Commission & NISA (around 3:20)	
4:00			Notify ministries etc. of insufficient depressurization in Unit 2	PM & TEPCO CEO meeting Agreed to set up integrated government/TEPCO response HQ (4:17)	

TIME	DISASTER/ACCIDENT	FUKUSHIMA DAIICHI	TEPCO	METI/NISA	PM/KANTEI
5:00					PM announces set up of integrated government/TEPCO response HQ (5:26) PM arrives at TEPCO Head Office. Stays in Integrated Response HQ (5:35–)
6:00	Loud impact sound in vicinity of Unit 2 pressure control chamber; Unit 4 building damaged (around 6:10)				PM delivers approx. 10-min long speech at TEPCO Integrated Response HQ
7:00		Contact ministries etc. of temporary withdrawal to Fukushima Daini excepting supervisory and operations staff (7:00)	Notify NISA of damage to Unit 4 building (7:55)		
8:00					PM leaves Integrated Response HQ (8:39) PM arrives at Kantei (8:46)
10:00				METI Minister orders early water injection into Unit 2 reactor, and venting if required, and fire extinguishing and avoidance of recriticality in Unit 4 (10:36)	
11:00					Indoor evacuation order for residents within 20–30km radius of Fukushima Daiichi NPS (11:00)
15:00					PM instructs Minister of Defense etc. that water injection required (15:58)

MAJOR EVENTS FROM MARCH 16

- **Fire outbreak in Unit 4 Building (early hours of March 16)**
- Unit 3 Trouble (March 16–23):
  March 16　two eruptions of white smoke in the morning Central Control Room operators withdraw temporarily; return at 11:33
  March 20　temperature in the reactor exceeds 300 degrees; pressure also climbing
  March 23　16:20　Black smoke eruption
  　　　　　23:30　Confirm smoke has ceased
- **Water injection/spraying into spent fuel pools (March 16–)**
- N.B. See below for details
- Water accumulation in the pit exceeding 1,000mSv/h (April 2)
- Ocean water discharge of low-level accumulated water in the centralized waste treatment facility (April 4–10)

EQUIPMENT USED FOR WATER INJECTION INTO SPENT FUEL POOLS

March 16	p.m.		Atmospheric dose high; forego water spraying by GSDF helicopter
March 17	a.m.	UNIT3	GSDF helicopter
	p.m.	UNIT3	Police water cannon truck
		UNIT3	SDF fire truck
March 18	p.m.	UNIT3	SDF fire truck
		UNIT3	US high-pressure water cannon truck (by TEPCO staff)
March 19			
	late evening	UNIT3	(Tokyo Fire Department) Hyper Rescue Squad fire truck
	p.m.	UNIT3	Hyper Rescue Squad fire truck
March 20	a.m.	UNIT4	10 SDF fire trucks
	p.m.	UNIT3	10 SDF fire trucks
	night	UNIT4	Hyper Rescue Squad fire truck
March 21	a.m.	UNIT4	12 SDF fire trucks & 1 US high-pressure water cannon truck
March 22	p.m.	UNIT3	Hyper Rescue Squad fire truck
After March 22		UNIT4	Concrete pump truck (Giraffe)
After March 27		UNIT3	Concrete pump truck (Giraffe)
After March 31		UNIT1	Concrete pump truck (Giraffe)

一般財団法人　日本再建イニシアティブ
民間事故調報告書検証チーム

Anatomy of the Yoshida Testimony
—— The Fukushima Nuclear Crisis
　　as seen through the Yoshida Hearings

発　行　日	2015年8月28日　第1刷発行
著　　　者	日本再建イニシアティブ 民間事故調報告書検証チーム 一般財団法人　日本再建イニシアティブ 〒107-0052　東京都港区赤坂2-23-1 アークヒルズ　フロントタワー　RoP 11階 電話　03-5545-6733 FAX　03-5545-6744
発　行　者	田辺修三
発　行　所	東洋出版株式会社 〒112-0014　東京都文京区関口1-23-6 電話　03-5261-1004（代） 振替　00110-2-175030 http://www.toyo-shuppan.com/
アートディレクション	熊澤正人
デザイン・レイアウト	大谷昌稔、村奈諒佳（パワーハウス）
イラストレーション	丸山幸子
校　　　正	丸山カオリ（円水社）
印刷・製本	日本ハイコム株式会社

許可なく複製転載すること、または部分的にもコピーすることを禁じます。乱丁・落丁の場合は、
ご面倒ですが、小社までご送付ください。送料小社負担にてお取り替えいたします。
©一般財団法人 日本再建イニシアティブ, 2015, Printed in Japan.
ISBN978-4-8096-7801-1
定価はカバーに表示してあります
ISO14001 取得工場で印刷しました